CONTENTS

PREFACE

This edition of *Problems in Engineering Graphics* by Arnbal and Crawford has been designed to be used in conjunction with a basic graphics course for freshman engineering students.

Sufficient, appropriate material in both range and degree of difficulty has been included to fully meet the requirements of either a 3-4 quarter hour credit, or a 2-4 semester hour credit course, covering fundamentals in what is commonly referred to as basic graphics or drawing and descriptive geometry.

In perusing the material in this book it is suggested that the following points should be thoughtfully noted.

• The table of contents lists easy to find, easy to assign major subject categories stated in commonly used terminology.

• Each major unit is introduced with a statement of objectives for the problems involved and a brief explanation of the relevance of the problems in both the educational and industrial environments.

• Solution principles are stressed prior to, and in conjunction with, the application problems.

• Special effort has been made to arrange problems for a given unit in a carefully graduated sequence starting with the simplest and progressing to the more challenging application problems.

• In each new unit introduced, deliberate review of preceding principles is incorporated in the solution requirements.

This problem book and the text *Engineering Graphics* by DeJong, Rising and Almfeldt is used in a sequence of engineering courses required in all engineering curricula.

Although subject areas and certain terms in this book have been "keyed" to the text *Engineering Graphics*—DeJong, Rising and Almfeldt the selection, grouping of material and the terminology used will be found compatible with any of the graphics texts currently in publication.

The authors hope that students using the problem sheets herein will appreciate having the repetitive, tedious and time consuming layout work of the respective problems accomplished for them in order that they may achieve a more interesting and comprehensive understanding of important principles and techniques than would otherwise be possible.

The authors would be pleased to receive both constructive criticism and enthusiastic endorsements from students and others using this edition.

FREEHAND DRAWING

LEARNING GOALS

The student should:

- Become familiar with the various drawing systems that are generally used in making freehand drawings.

- Be able to make a logical selection of drawing systems when planning a freehand drawing of a specific object.

- Be able to set up a freehand drawing to the best advantage with respect to size and orientation.

- Develop an ability to visualize proportions in his mind and transfer that information into a well proportioned drawing.

- Develop sufficient skill in making freehand lines to facilitate the transfer of information from the mind into an aesthetically pleasing freehand drawing.

APPLICATIONS

- Recent developments in the engineering profession have made the practicing engineer more dependent on freehand drawing and less on instrument drawing for communication with colleagues and others. Freehand drawing is used to facilitate retention, stimulate creativity, develop concepts, and communicate in a wide variety of situations.

1. SKETCHING—EXECUTION

SKETCHING

A. 1. Keeping your eyes focused on point B (the objective) sketch a straight line from A to B, then from C to D, C to A, and D to B.
2. Divide the rectangle ABCD into 6 equal areas as follows. By estimating, freehand, divide line AB into three equal spaces, and AC into two equal spaces. Sketch lines parallel to AC and AB respectively.

A +   + B

C +   + D

B. 1. Very lightly sketch lines through points X, Y, and Z that are parallel to line PQ.
2. Very lightly sketch lines through points X, Y, and Z that are perpendicular to line PQ.
3. Check positions of the lines sketched in 1 and 2 above. If the parallelism and perpendicularity are satisfactory, darken the lines to a finished object line quality.

C. 1. Sketch a line that divides MNO into two 45° angles.
2. Sketch two lines that divide RST into three 30° angles.
3. Sketch a square using EF as one side.
4. Sketch a circle inscribed in the given square.
5. Using JK as one side sketch a 30°-60°-90° triangle. (Check angles after you have finished.)

1. M   2. R   3. E

N   O   S   T   F

4.   5.

J ———————————————— K

DR. No.
GRADE:
DR. BY:
INSTRUCTOR:
DATE:
TABLE:
COURSE:
SECTION:
2

2. SKETCHING—PICTORIAL VIEWS

A. Study each object and select the view that best shows the contours. Make a single view sketch of each utilizing the space provided.

1.	2.	3.

4.	5.	6.

1.	2.	3.	4.	5.	6.

COURSE : ‾
SECTION : ‾
DATE : ‾
TABLE : ‾
DR. BY : ‾
INSTRUCTOR : ‾
GRADE :

3

3. SKETCHING—FROM PHOTOGRAPHS

A. Make freehand single view sketches as assigned.

B. Make freehand pictorial sketches as assigned.

Dr. No.:

Grade:

Dr. By:

Instructor:

Date:

Table:

Course:

Section:

Courtesy of Electonic Products
Division E G & G Inc.

1. Radiometer detector head

Courtesy of International Crystal
Mfg. Company

4. Switch

Courtesy of Electronic Products
Division

2. Variable resistor

Courtesy of Aluminum Company of America

5. Conductor clamp

Courtesy of Bijur
Lubricating Corp.

3. Pilot locator

Courtesy of Koehring Company

6. Roller

4. SKETCHING—MEMORY AND SEQUENCE INSTRUCTIONS

SKETCHES

A. From the following titles, select subjects for sketches. Proportion the sketch to fit the working area on the sheet.

1. Baseball bat	8. Carpenter's saw	15. Study desk
2. Screwdriver	9. Traffic light	16. Scotch tape dispenser
3. Electric light bulb	10. Coffee pot	17. Floor plan of home
4. Automobile instrument panel	11. Wall clock	18. Desk lamp
5. Wrist watch	12. Bicycle	19. Guitar
6. Paper clip	13. Fluorescent light	20. Transistor radio
7. Water tower	14. Automobile	

B. Pictorial sketch.

Starting with very light construction lines and using symmetrical axes, make a pictorial sketch conforming to the following sequence of specifications:

1. Sketch a rectangular prism 8 units long (receding to the left), 4 units high and 6 units deep.

2. Locate a line around the prism at a height of 2 units.

3. Locate a line around the prism 4 units to left of the front corner (your construction will not look like 4 blocks each 4 × 2 × 6).

4. Remove the entire 4 × 2 × 6 block from the upper front corner.

5. On the the remaining upper block, locate a line joining the midpoints of the 4 unit distance.

6. Now cut two triangular prisms from this block by drawing lines from each of the lines located in step 5 down to the lower corners of the upper block.

7. Remove a 2 × 2 × 4 portion from the front rectangular block to leave a 4 × 2 × 4 remainder.

8. Add door, windows, steps, sidewalk, etc., to transform your mutilated block into a house with an attached garage.

C. Multi-view sketch.

Starting with very light construction lines, make a sketch conforming to the following instructions:

1. Make a two view sketch (top and front) of a rectangular prism that is 8 units long, 4 units high and 6 units deep. Front view is 8 units wide and 4 units high. Top view, placed above the front view is 8 units wide and 6 units deep.

2. Divide the front view in half with a horizontal line. (Does not change top view)

3. Divide top and front views in half with a vertical line. (Divides prism into four blocks, two visible from the top and all four visible from front.)

4. Remove the upper right block in the front view. (Does not change top view, although the right lower block is now visible from the top.)

5. Locate a point in the middle, two units from either side, of the upper line of the upper left hand block in the front view. From this point, draw 45° lines to each of the lower corners of the upper left hand block.

6. In the top view, draw a vertical line in the middle, two units from either side, of the left hand block.

7. In the front view, remove the triangles above the 45° lines in the upper left block. (Does not change the top view although slanted surfaces are now visible on the left side of the top view.)

8. In the top view, remove the lower two units of the right hand block. (Leaves a 4 × 4 block, two units high on the right side.)

You now have top and front elevation views of the house drawn in pictorial in problem "B" above.

DR. NO.:__ GRADE:__ DR. BY:__ INSTRUCTOR:__ DATE:__ TABLE:__ COURSE:__ SECTION:__

ENGINEERING LETTERING

LEARNING GOALS

The student should:

- Comprehend basic practices that result in good lettering. Included would be the use of guide lines to maintain uniform size, recognition of basic gothic letter shapes for maximum legibility, and constant letter slope for uniform appearance. Standard practices for spacing contribute heavily to good readability.

- Develop a degree of skill in lettering such that his lettering will not only be easily legible, but will also contribute to the good appearance of the drawing.

- Develop a rate of speed in lettering that is as rapid as possible. This might be the most important goal to be considered in this area.

APPLICATIONS

- Lettering is a form of graphical communication with almost unlimited application. Nearly every drawing, graph, chart, or any other type of graphical material requires some written material to make it complete, and in many situations this material is hand lettered. Additional applications involve engineering computations, reports, record keeping, and many others.

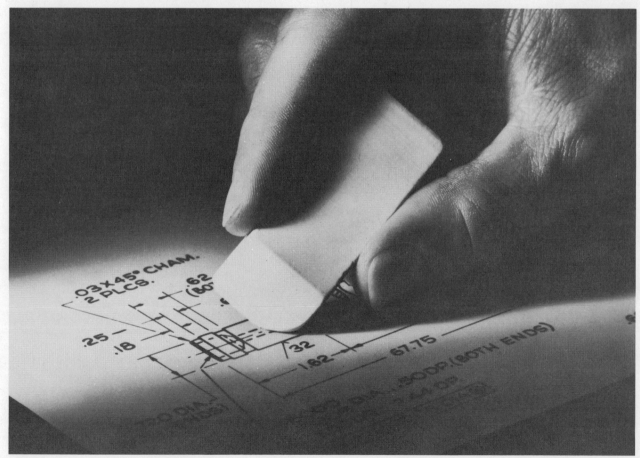

Eastman Kodak Company

5. STANDARD LETTERS AND NUMERAL CHART

ABCDEFGHIJ
KLMNOPQRST
UVWXYZ&8KR
abcdefghijklmn
opqrstuvwxyz
1234567890

ABCDEFGHIJ
KLMNOPQRST
UVWXYZ&8KR
abcdefghijklmn
opqrstuvwxyz
1234567890

7

DR. No.___
GRADE:

COURSE: 165   DATE: 9-15-   DR. BY: JOHN JOHNSON
SECTION: 3B   TABLE: 14   INSTRUCTOR: M.W. ALMFELDT

IOWA STATE UNIVERSITY
LETTERS AND NUMERALS

6. LETTERING EXERCISE: SHAPE CHARACTERISTICS

LETTERING EXERCISES

A. Using the grids and squares as guides, letter the alphabet in vertical caps in two letter sizes as indicated.

B. Using the grids and squares as guides letter the vertical numerals in two letter sizes.

C. Using the squares on the guide lines letter the alphabet in lower case letters.

D. On the guide lines below letter in vertical caps the following sentence: "Uniform height and proper spacing contribute heavily to the appearance of engineering lettering."

7. LETTERING EXERCISE: USE OF GUIDE LINES

(Make guide lines with instruments or freehand as assigned)

A. Make very light guide lines for 3 mm letter height and carefully letter in the missing words in the following statements (each will be one of the five S's of lettering).

Good lettering results from: learning the characteristics in form and proportion (the

_____ of each letter or nu-

meral); consistently using guide lines to insure uniformity in the _____

of each character; attention to the uniform position of the characters in relation to a

theoretical axis (the _____ of the letter) (in this respect, lettering is classified

as either _____ or _____); always main-

taining a proper void area (_____) between letters within a word

or between words within a sentence; persistent attention to the foregoing features together

with diligent effort and practice and you will be able to letter at a satisfactory rate of

_____ .

B. Make very light guide lines for 4 mm letter height and letter in vertical caps: your curriculum, your advisors name, your home town, and your social security number.

C. On the given electrical layout identify each symbol by carefully lettering its name by the appropriate leader. Make very light guide lines similar to those that have been drawn for the fuse. The symbols indicate the following components:

   1. Fuse
   2. Transformer
   3. Diode
   4. Switch
   5. Capacitor
   6. Resistor

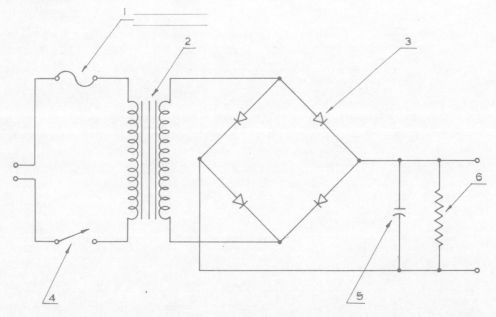

8. LETTERING EXERCISE: APPLICATION

A. Letter the following dimensions and notes where indicated. Use guide lines for 3 mm letter height and letter in vertical caps.

1. 13.0
2. 6.0
3. 15.0
4. 8.5
5. 12.5
6. 25.0
7. 5 mm drill
   2 holes
8. 7 mm drill
   2 holes
9. 15.0
10. 12.5
11. 76.0
12. 35.0
13. 2.00
14. Note:
    Hardness: Rockwell
    C40–C45
15. Note:
    CAD Plate .008 mm thick
    Heat treat to remove hy-
    drogen embrittlement
16. Title Plate:
    Name: Friction Spring
    Material: 1065–1085 car-
    bon spring steel 2.00 × 25
    × 112 mm long
    No. Reqd: 2
    Scale: 1:1

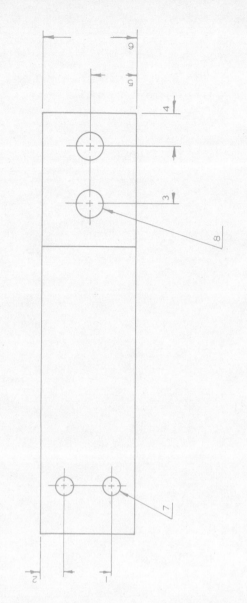

16. TITLE PLATE :

14. NOTE :

15. NOTE :

COURSE : ⁻	DATE : ⁼
SECTION : ⁻	TABLE : ⁼
	DR. BY : ⁼
	INSTRUCTOR : ⁼
	DR. NO. ⁻
	GRADE :

10

DRAWING EQUIPMENT

LEARNING GOALS

The student should:

- Learn to use his drawing equipment in an efficient manner. Demonstrate the accepted standard techniques in using compass, scales, triangles and other drawing equipment.

- Arrange equipment on his desk for orderly use. A place for everything and everything in its place.

APPLICATIONS

- Equipment is used whenever an accurate, scale drawing is desired of any conceived design.

Martin Instrument Company

9. DRAWING EQUIPMENT: INSTRUMENTS APPLICATION

Instrument exercises—notice line weights and form.

A. Redraw each figure in space to its right.

B. Make double size drawing of each figure on a sheet of blank paper. Use dividers to transfer measurements.

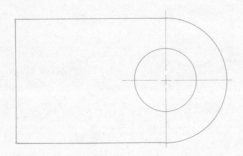

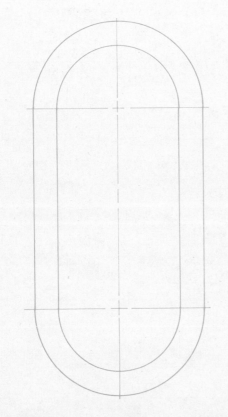

10. DRAWING EQUIPMENT: ENGINEERING SCALE (ENGLISH SYSTEM)

USE OF ENGINEERS SCALE

Draw appropriate guidelines and record each missing scale, ratio, dimension (value and unit), and physical measurement (arrowhead).

A. Do all 20 exercises.

B. Complete only the odd numbered exercises.

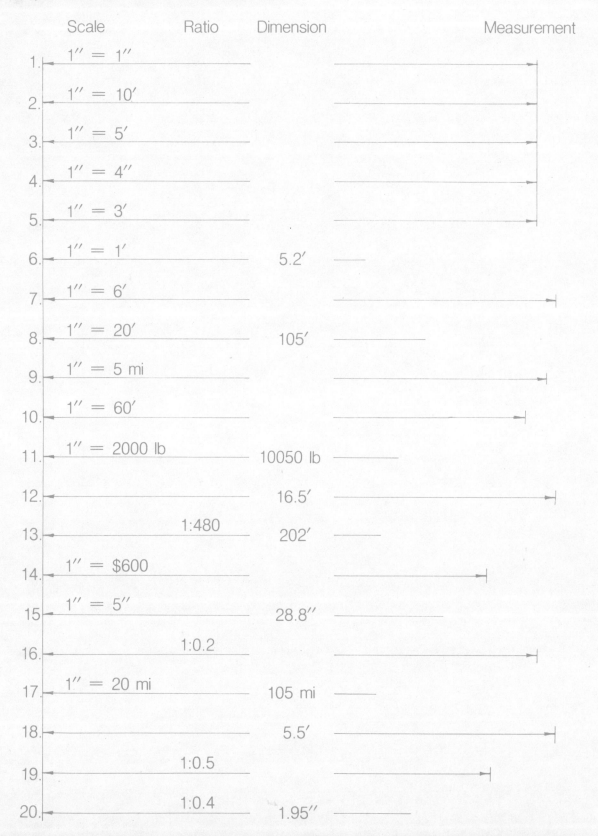

	Scale	Ratio	Dimension	Measurement
1.	1″ = 1″			
2.	1″ = 10′			
3.	1″ = 5′			
4.	1″ = 4″			
5.	1″ = 3′			
6.	1″ = 1′		5.2′	
7.	1″ = 6′			
8.	1″ = 20′		105′	
9.	1″ = 5 mi			
10.	1″ = 60′			
11.	1″ = 2000 lb		10050 lb	
12.			16.5′	
13.		1:480	202′	
14.	1″ = $600			
15.	1″ = 5″		28.8″	
16.		1:0.2		
17.	1″ = 20 mi		105 mi	
18.			5.5′	
19.		1:0.5		
20.		1:0.4	1.95″	

DR. NO.:__ GRADE:__

DR. BY:__ INSTRUCTOR:__

DATE:__ TABLE:__

COURSE:__ SECTION:__

11. DRAWING EQUIPMENT: ENGINEERING SCALE (METRIC SYSTEM)

Draw appropriate guidelines and record each missing ratio, dimension and physical measurement (arrowhead).

A. Do all 20 exercises.

B. Do only the odd numbered exercises.

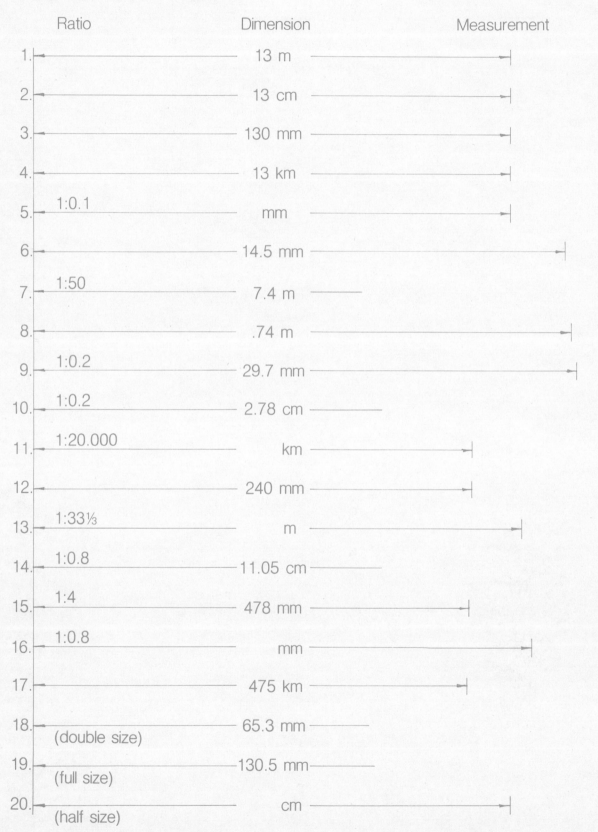

	Ratio	Dimension	Measurement
1.		13 m	
2.		13 cm	
3.		130 mm	
4.		13 km	
5.	1:0.1	mm	
6.		14.5 mm	
7.	1:50	7.4 m	
8.		.74 m	
9.	1:0.2	29.7 mm	
10.	1:0.2	2.78 cm	
11.	1:20.000	km	
12.		240 mm	
13.	1:33⅓	m	
14.	1:0.8	11.05 cm	
15.	1:4	478 mm	
16.	1:0.8	mm	
17.		475 km	
18.	(double size)	65.3 mm	
19.	(full size)	130.5 mm	
20.	(half size)	cm	

DR. NO.:
GRADE:

DR. BY:
INSTRUCTOR:

DATE:
TABLE:

COURSE:
SECTION:

ORTHOGRAPHIC PROJECTION

LEARNING GOALS

The student should:

- Demonstrate an understanding of the concepts of orthographic projection; planes of projection and their relationship to the viewer, and to the object.

- Demonstrate skill and proficiency in the mechanics of projecting; perpendicularity, parallelism, measurements and line quality when preparing drawings with instruments or by sketching.

APPLICATIONS

- Orthographic projection is used world wide as a system of drawing whenever accurate and detailed information is to be transmitted regarding the shape, size, manufacture, assembly or function of any product or design.

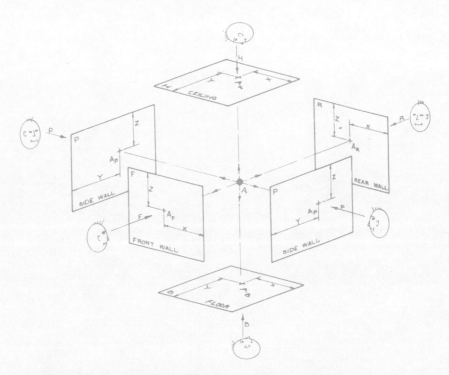

PRINCIPAL PLANES OF PROJECTION

16

12. THEORY OF ORTHOGRAPHIC PROJECTION
13. ORTHOGRAPHIC PROJECTION—HINGE LINE MEANINGS

ORTHOGRAPHIC PROJECTION

1. Theory of Orthographic Projection
   Your ability to graphically communicate and solve problems is dependent upon a clear understanding of the principles of orthographic projection. Study your text carefully in this topic area.

   A. At the appropriate places on the given sketch, letter the "four significant features" of orthographic projection.

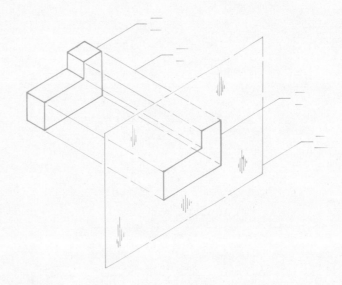

   B. A definition of orthographic projection contains the above four significant features. In your own words, write the definition.

2. Understanding the meaning of the hinge line (———————— — — ————————) is necessary to the understanding of the system of orthographic projection. The correct interpretation of any hinge line is dependent upon the position of the viewer in relation to the planes of projection.

   List two different interpretations of what each of the following hinge lines represents. Describe clearly the position of the viewer or point of view necessary to make each interpretation valid.

   I. ——————— – – – $\frac{H}{F}$ ———————

   2. F^I

   l. a.

   b.

   2. a.

   b.

14. ORTHOGRAPHIC PROJECTIONS—POINTS

14a. PROBLEM: PRINCIPAL VIEWS
14b. PROBLEM: PRINCIPAL VIEWS

A. Application Problem—Points
   Ratio 1:0.8

   Locate H, F & P projections of
   point A which is:

       30mm back of F
       50mm below H
       40mm left of P

   & B which is: 36mm back of F
               42mm below H
               27mm left of P

B. Application Problem—Points
   Ratio 1:40

   1. Locate wall outlet "O" which
      is on the side wall (P).

      1.30m below the ceiling (H)
      1.50m back of front wall (F)

   2. Find ceiling connection "C"
      which is:

      on the ceiling (H)
      1.8m L of side wall (P)
      2.5m back of F wall (F)

17

DR. NO.__
GRADE:

DR. BY:__
INSTRUCTOR:__

COURSE:__
SECTION:__

DATE:__
TABLE:__

14c. PROBLEM: AUXILIARY VIEWS
14d. PROBLEM: AUXILIARY VIEWS

C. 1. Locate and label the projections of point A in all of the given views. (Ratio 1:50, measured in meters.)

2. Point B is located 2 below H, 0.5 behind F and 1.62 to left of P. Project point B into all of the given views and label.

3. Point C is located 2 below H, 1 behind F and 0.37 to left of P. Project point C into all of the given views and label.

D. 1. Locate and label the projections of points D and E in all of the given views.

2. Point L is located on the bottom rim of the cylinder, 50mm behind F and 115mm to left of P. Project point L into all of the given views and label. (Ratio 1:4)

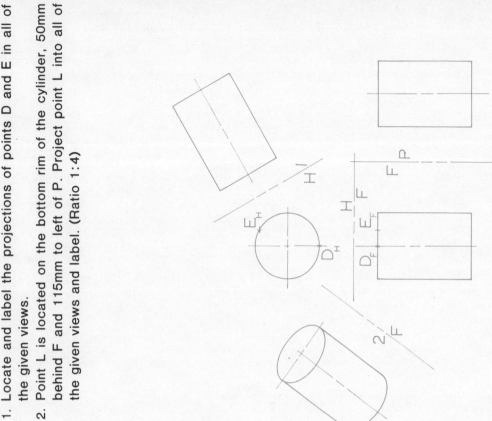

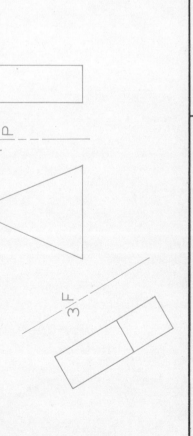

18

COURSE: ⁼_   DATE: ⁼_   DR. NO.⁻_
SECTION:⁻_   TABLE:⁻_   DR. BY: ⁼_
                              INSTRUCTOR: ⁼_
                              GRADE:

14e. PROBLEM: PROJECTION OF POINTS TO AUXILIARY VIEWS

ORTHOGRAPHIC PROJECTION—POINTS

1. Freehand (estimate distances):

2. Instruments: (Ratio 1:1, All distances in millimeters)

Show the indicated projections of point A located 25 behind the front plane, 15 below the horizontal plane, and 55 to the left of the profile plane.

Point C is located 10 behind the front plane, 30 below the horizontal plane and 20 to the left of the profile plane.

DR. NO.:
GRADE:

DR. BY:
INSTRUCTOR:

DATE:
TABLE:

COURSE:
SECTION:

15. ORTHOGRAPHIC PROJECTION—LINES

LEARNING GOALS

The student should:

- Develop an ability to visualize lines in various positions.

- Be able to apply the theory of orthographic projection to find specified views of lines.

- Be able to find and measure line specifications from given orthographic views.

- Be able to delineate a line from given specifications.

- Be able to classify lines according to position.

APPLICATIONS

- Most engineering drawings are comprised of combinations of lines that may or may not be in defined positions.

- Graphical representation of data, graphical mathematics, and vector analysis depend on the positions and characteristics of lines.

- Problems involving clearances, intersections, and relative positions of structural members, wires, pipes, highways, etc. require a knowledge of lines for graphical solution.

Owens-Corning Fiberglas Corporation

15a. PROBLEM: CLASSIFICATION

DR. NO.__ GRADE:

1. The seven basic positions for lines are illustrated. Remember the actual line is never seen. Each line is delineated by orthographic images placed in projection.

1.	2.	3.	4.	5.	6.	7.
A_H B_H H/F A_F B_F	C_H D_H H/F D_F C_F	E_H F_H H/F E_F F_F	G_H H_H H/F H_F G_F	K_H J_H H/F J_F K_F	$L_H M_H$ $+$ H/F L_F M_F	P_H O_H H/F O_F P_F

2. The following statements concern the lines illustrated in "1" above. Use double classifications when applicable.

  1. Line AB is parallel to the _____ plane.

  2. Parallelism between line AB and the _____ plane is evident from the position of the _____ projection.

  3. The classification of line AB is _____ .

  4. Line CD is parallel to the _____ plane.

  5. Parallelism between line CD and the _____ plane is evident from the position of the _____ projection.

  6. The classification of line CD is _____ .

  7. Line EF is parallel to the _____ plane.

  8. Parallelism between line EF and the _____ plane is evident from the position of the _____ projection.

  9. The classification of line EF is _____ .

  10. Line GH is parallel to the _____ plane.

  11. Parallelism between line GH and the _____ plane is evident from the position of the _____ projection.

  12. The classification of line GH is _____ .

  13. Line JK is parallel to the _____ plane.

  14. Parallelism between line JK and the _____ plane is evident from the position of the _____ projection.

  15. The classification of line JK is _____ .

  16. Line LM is parallel to the _____ plane.

  17. Parallelism between line LM and the _____ plane is evident from the position of the _____ projection.

  18. The classification of line LM is _____ .

  19. Line OP is parallel to the _____ plane.

  20. Parallelism between line OP and the _____ plane is evident from the position of the _____ projection.

  21. The classification of line OP is _____ .

DR. BY:__ INSTRUCTOR:__ DATE:__ TABLE:__ COURSE:__ SECTION:__

15b. PROBLEM: TRUE LENGTH AND POINT OF VIEW

1. Solution Principles

    1. A line appears in true length when projected on any plane that is parallel to the line.

    2. A line appears as a point when projected on any plane that is perpendicular to the line. (Perpendicular to a T.L. projection)

2. Determine and label the true length (TL) and point view (PV) projections of each of the following lines.

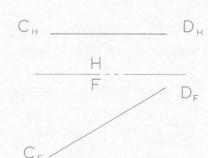

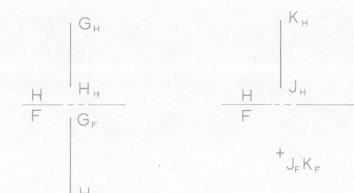

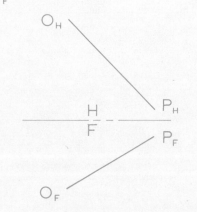

3. Classify each of the above lines

    AB _____

    CD _____

    EF _____

    GH _____

    JK _____

    LM _____

    OP _____

DR. NO.__

GRADE:

DR. BY:__

INSTRUCTOR:__

DATE:__

TABLE:__

COURSE:__

SECTION:__

15c. PROBLEM: LINES—BEARING

1. Solution Principle

   The bearing of a line is always determined in the horizontal projection. The bearing angle is the acute angle between the horizontal projection and a N-S line through the origin.

2. Determine and label the bearing angle and bearing for each of the following lines.

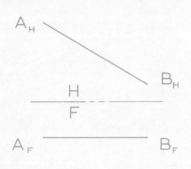

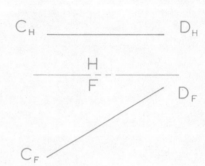

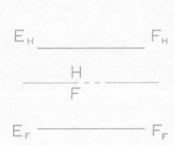

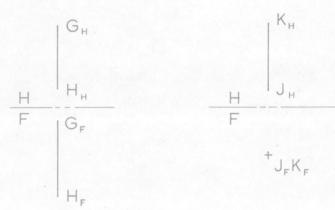

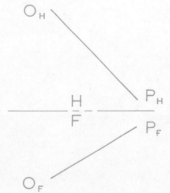

3. Classify the plane of projection in which each of the following is found.

LINE	BEARING	T.L.	P.V.
AB			
CD			
EF			
GH			
JK			
OP			

15d. PROBLEM: LINES—INCLINATION

1. Solution Principle

   The inclination of a line is in true size only on a plane of projection that is an elevation plane and is parallel to the line. It is measured from the edge view of a horizontal plane through the origin of the line to the TL projection of the line. The form may be an angle (slope angle) or a ratio of $\dfrac{\text{vertical}}{\text{horizontal}} = \dfrac{\text{rise}}{\text{run}}$ (slope or grade).

2. Determine and label the slope angle of the following lines.

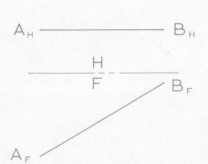

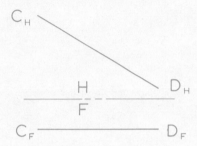

3. Determine the slope of the following lines. Label $\dfrac{\text{rise}}{\text{run}}$ in the solution view.

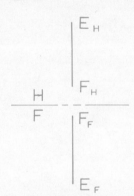

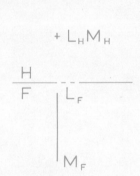

4. Determine the grade of the following lines. Label $\dfrac{\text{rise}}{\text{run}}$ in the solution view.

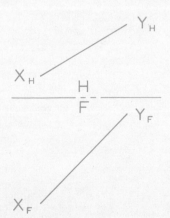

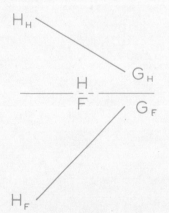

DR. No.—
GRADE:

DR. BY:—
INSTRUCTOR:—

DATE:—
TABLE:—

COURSE:—
SECTION:—

15e. PROBLEM: MEASUREMENT OF LINE SPECIFICATION

1. Each specification must be measured and labeled in the projection that satisfies the solution principle.

2. Determine and label the true length, point view, bearing, and inclination (slope angle for AB, slope for CD, grade for XY) for each line.

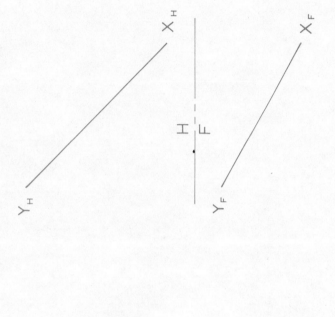

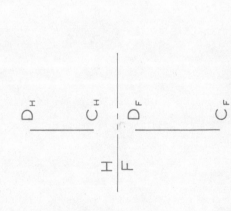

3. Classify each line: AB _____, CD _____,
XY _____.

| COURSE: | DATE: | DR. BY: |
| SECTION: | TABLE: | INSTRUCTOR: |

DR. NO.
GRADE:

25

15f. PROBLEM: DELINEATE FROM SPECIFICATIONS

1. Delineate the following lines from given specifications and complete the H and F projections. Label in the solution projection each specification used.

LINE	BEARING	INCLINATION	T.L.	SCALE
AB	Due W	+30° SA	17 cm	1:5
CD	Due S	−1.2 slope	35 mm	1:1
XY	N60° E	+75% Grade	40 m	1:800

27

15g. PROBLEM: PRACTICAL APPLICATION (TRUE LENGTH, BEARING AND SLOPE ANGLE)

DR. No.___ GRADE:

DR. BY:≡ INSTRUCTOR:≡

DATE:≡ TABLE:≡

COURSE:⁼ SECTION:⁼

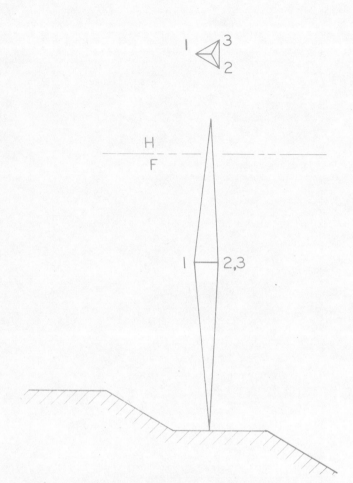

Given: H and F projections of a tower
with ground planes shown in
the front projection. (Scale: 1″ = 40′)

1. Find: Length of wire needed for three equally spaced guy wires at (−)45° slope angles from the corners of the structure to the ground.

2. Bearings of the wires will be (1) _____
   (2) _____ (3) _____

3. The slope angle for guy wire 1 can be shown in T.S. in _____ projection.

4. The slope angle for guy wire 2 (or 3) can be shown in T.S. in _____ projection.

5. The total length of wire to reach the ground is _____ .

Note: When the position for the lines has been established graphically the points where the lines pierce the ground planes can be found by extending the front projections.

Courtesy of Reynolds Metals Company

28

15h. PROBLEM: PRACTICAL APPLICATION—BEARING, GRADE, TRUE LENGTH

1. To solve the following problems you must use correct principles of orthographic projection in conjunction with the solution principles for line specifications.

2. Delineate each line by determining H and F projections plus any other projections that are necessary for the information to complete the table. Label each specification in this solution view. Scale 1:1 in mm.

LINE	CLASSIFICATION	TL	BEARING	GRADE (%)
AB				
CD				
RS				
XY				

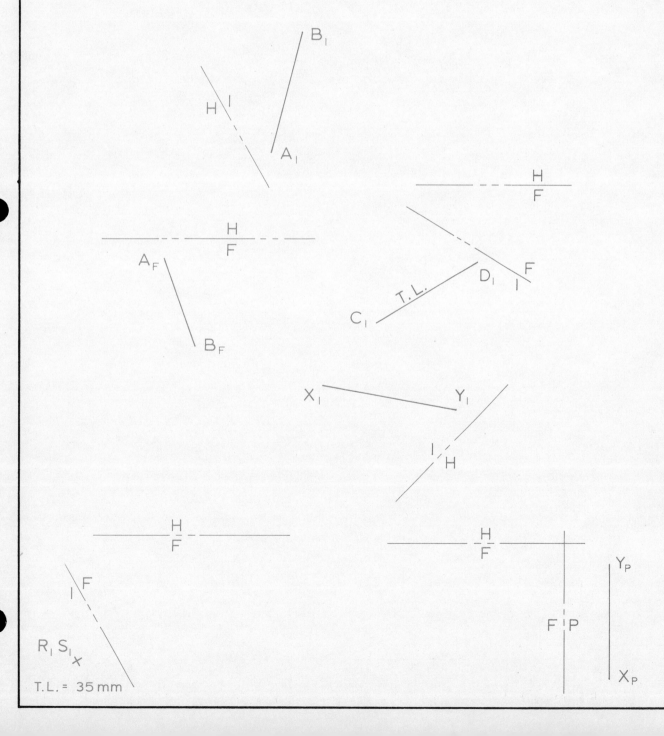

T.L. = 35 mm

DR. NO._

GRADE:

DR. BY:—
INSTRUCTOR:—

DATE:—
TABLE:—

COURSE:—
SECTION:—

29

16. ORTHOGRAPHIC PROJECTION—PLANES

LEARNING GOALS

The student should:

- Be able to graphically represent a plane in its various forms.

- Be able to solve problems including edge view, inclination and true size of a plane.

- Be able to locate points or lines on the surface of a plane.

- Be able to visualize the orientation of a plane surface in three dimensional space.

APPLICATIONS

- Of all types of surfaces that exist, the plane surface is the one most frequently used in all types of design. The basic principles involved with planes are applicable in many design areas.

Maho Machine Tool Corporation

16a. PROBLEM: CLASSIFICATION

1. Classify each of the planes. Mark T.S. on views showing the true size projection of the plane.

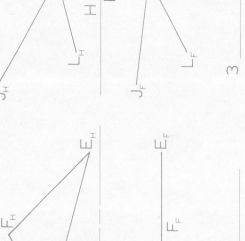

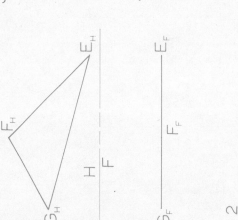

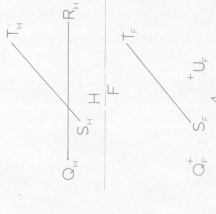

2. Complete (freehand) the views of the planes. All indicated points are on the plane in each exercise.

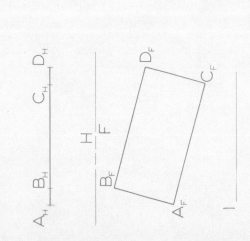

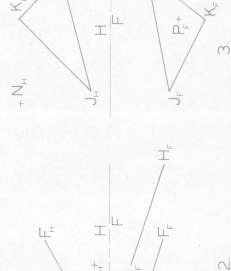

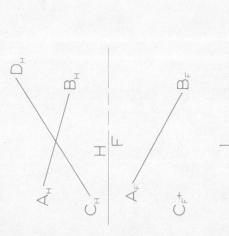

DR. NO.	
GRADE:	

30

COURSE:	DATE:	DR. BY:
SECTION:	TABLE:	INSTRUCTOR:

16b. PROBLEM: TRUE SIZE

1. Find and label the true shape of each plane. Indicate the classification of each plane.

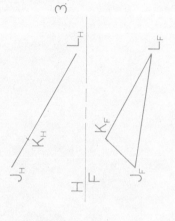

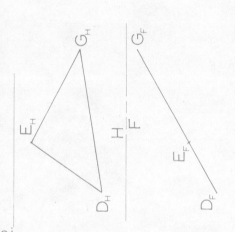

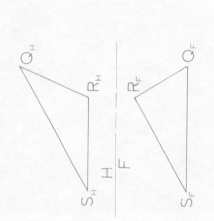

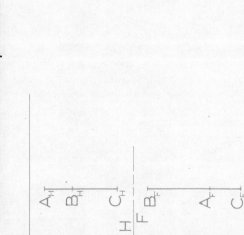

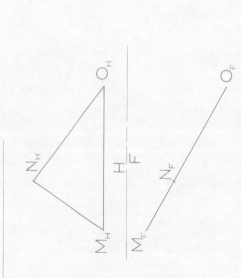

2. Write the solution principle for the true shape of a plane.

16c. PROBLEM: SLOPE ANGLE—TRUE SIZE—PRACTICAL APPLICATIONS

1. Sector ABCD is the surface of a steel plate, 10 mm thick, used in a construction project. Scale: 1:50.

   1. Determine the slope angle of this sector.

   2. Determine the weight of the plate. (Steel weighs 7860 kg/m³)

2. One segment of a chemical mixing tank is represented by plane sector MNOP. Scale: 1:10

   1. Determine the slope angle.

   2. Determine the area in square meters.

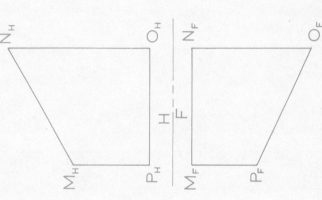

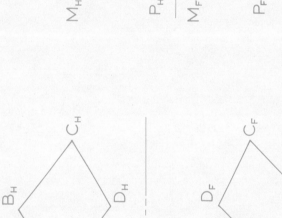

DR. NO.__

GRADE:

DR. BY:__

INSTRUCTOR:__

DATE:__

TABLE:__

COURSE:__

SECTION:__

16d. PROBLEM: SLOPE ANGLE—TRUE SIZE

1. Find the H and F views of
   the hexagon.

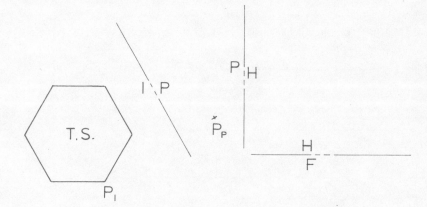

2. AB and CD lie in a plane
   whose slope angle is
   30°. AB is lower than CD.
   Find the TS of ABCD and
   project into front view.

3. Isoceles triangle (T.S.)
   XYZ appears as an edge
   in plane 3 with a slope
   angle of 30° sloping
   downward to the NW.
   Line XY is parallel to
   plane 3 and is 3 cm long.

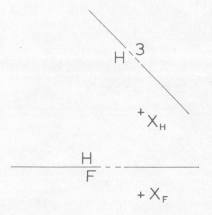

DR. BY:

INSTRUCTOR:

DATE:

TABLE:

COURSE:

SECTION:

34

17. SPACE GEOMETRY—POINTS AND LINES

LEARNING GOALS

The student should:

• Develop an ability to visualize lines and points in three dimensional space.

• Develop an ability to visualize a type of view that would be required to solve space and clearance problems involving points and lines.

• Develop an ability to perform the orthographic projections necessary to solve space and clearance problems.

• Develop a problem solving procedure.

APPLICATIONS

• The ability to identify shortest distances and clearances occurs any time objects must be placed in close proximity to each other. Orthographic projection is a valuable tool in the graphical determination of these values.

17a. PROBLEM: SHORTEST DISTANCE—POINT TO LINE

1. Locate the shortest connectors in all views.

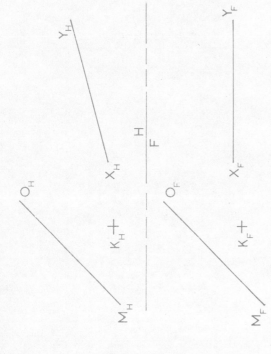

2. Determine which is shorter—the perpendicular connector from K to MO or the one from K to XY (Show connector in all views).

3. Locate the shortest connector in all views.

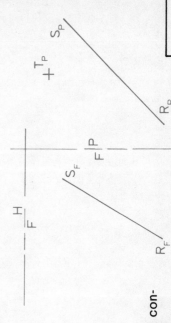

| COURSE: ‗ | DATE: ‗ | DR. BY: ‗ | DR. NO. ‾ |
| SECTION: ‗ | TABLE: ‗ | INSTRUCTOR: ‗ | GRADE: |

17b. PROBLEM: PLANE PARALLEL TO GIVEN LINE

1. Using just the H and F views, draw a plane ABC that contains line AB and is parallel to line DE. Prove your solution by an auxiliary elevation view.

2. Using just the H and F views, draw a plane XYZ containing point Z and parallel to both lines JF and GH. Prove your solution by an auxiliary elevation view.

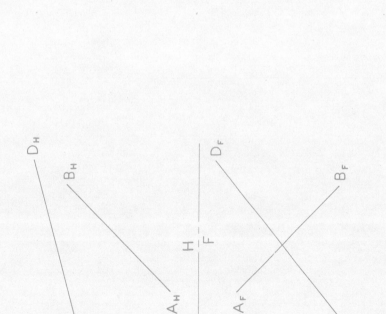

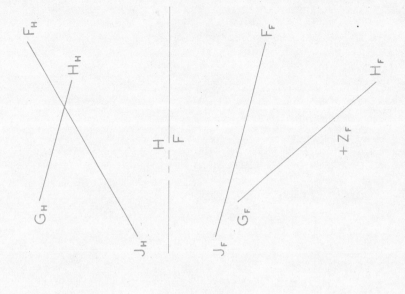

COURSE:	DR. BY:
SECTION:	DATE:
	TABLE:
	INSTRUCTOR:

DR. NO.:

GRADE:

17c. PROBLEM: SHORTEST CONNECTORS—PERPENDICULAR/VERTICAL

1. Show the vertical (if one is possible) and the perpendicular connectors in each view.

2. Classify, label and give the bearing and slope ∠ of each of the connectors found in 1.

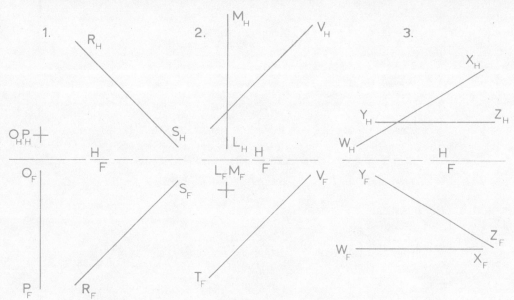

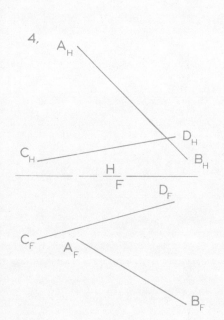

DR. NO.__

GRADE:

DR. BY:__

INSTRUCTOR:__

DATE:__

TABLE:__

COURSE:__

SECTION:__

17d. PROBLEM: SHORTEST CONNECTORS—VARIOUS

1. Show the vertical and perpendicular connectors in all views and label and record the classification, bearing and grade of each.

2. Show the shortest 50% grade, the shortest horizontal, the perpendicular and the vertical connectors in all views.

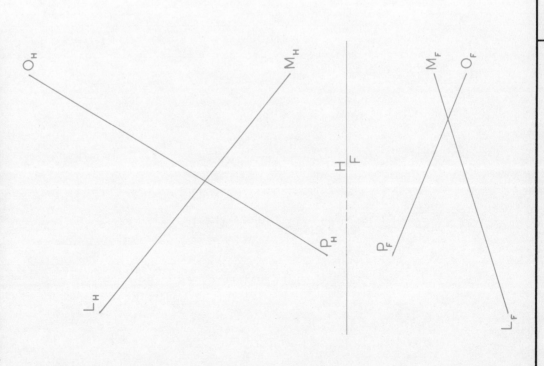

COURSE: ‾ DATE: ‾
SECTION: ‾ TABLE: ‾

DR. BY: ‾
INSTRUCTOR: ‾

DR. NO._
GRADE:

38

17e. PROBLEM: CLEARANCE INDUSTRIAL APPLICATION

Determine the clearance between the two pipes. Draw a line that would connect the two pipes at their closest point, and project into H and F views. Complete the H and F views with proper visibility. Scale: 1″ = 4″ Clearance = _____

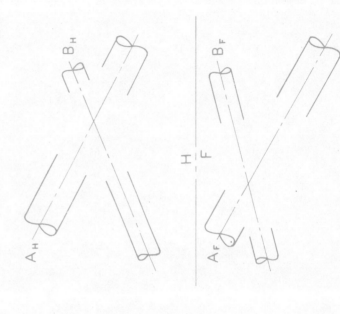

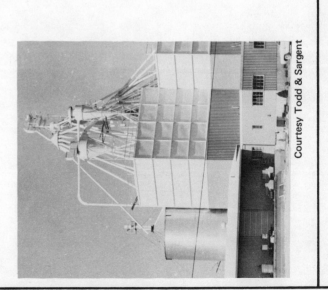

Courtesy Todd & Sargent

39

DR. NO.:___

GRADE:___

COURSE:___   DATE:___   DR. BY:___

SECTION:___   TABLE:___   INSTRUCTOR:___

17f. PROBLEM: CLEARANCE INDUSTRIAL APPLICATION

The location of a portion of a structural channel for a proposed hangar is shown. A level 3 × 3 × 0.4 angle iron strut has its corner passing through Point A bearing N75°E. Determine the lowest position of the angle which has one face parallel to the axis of the channel and passes above the channel with a one inch clearance. Scale: 1″ = 6″. Show the plan and elevation views of the angle iron with proper visibility.

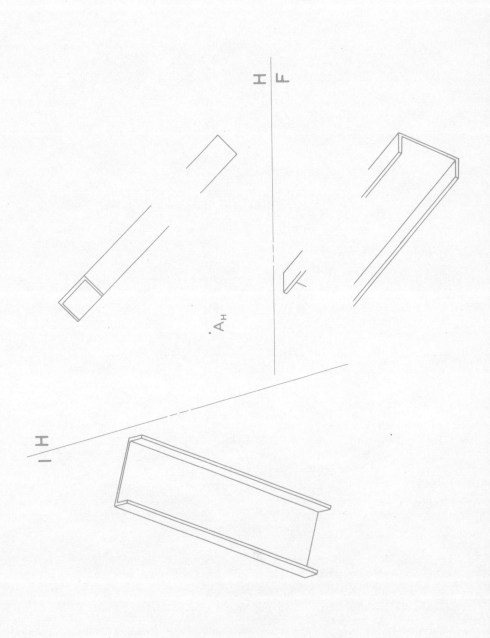

DR. No.‗

GRADE:‗

COURSE:═

SECTION:‗

DATE:═

TABLE:‗

DR. BY:═

INSTRUCTOR:═

18. SPACE GEOMETRY—LINES AND PLANES

LEARNING GOALS

The student should:

- Develop an ability to visualize various geometric shapes in three dimensional space.

- Develop an ability to analyze space geometry problems to select the solution principle that produces the most efficient solution.

- Develop an ability to perform the orthographic projections necessary to satisfy a solution principle.

APPLICATIONS

- Piercing point and intersection problems occur in many areas of engineering work.

- Angles between planes and between lines and planes occur when working with sheet metal, brackets, structural members, mineral strata, wiring, piping, and surfaces and bracing in many situations.

Terex Corporation

42

18a. PROBLEM: PIERCING POINT—LINE AND PLANE/EDGE VIEW METHOD

1. Solution Principle

   The intersection of a line and a plane is found in a projection showing the plane as an edge. Any adjacent view can be located by projection.

2. Use the edge view method to find point T where the line pierces the plane in each problem. Treat the planes as opaque and show proper visibility of each line.

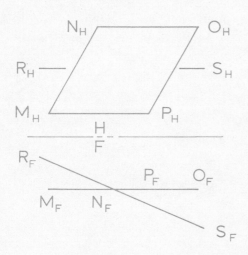

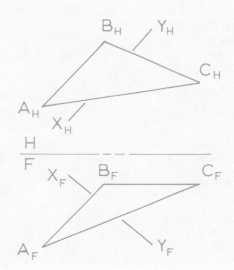

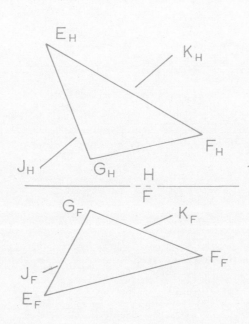

18b. PROBLEM: PIERCING POINT—LINE AND PLANE/CUTTING PLANE METHOD

43

1. Solution Principle

   An auxiliary cutting plane that contains the line forms a line of intersection with the given plane that intersects the given line at the piercing point.

2. Use the auxiliary cutting plane method to find point P where the line pierces the plane in each problem. Treat the planes as opaque and show the proper visibility of each line.

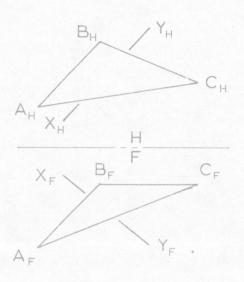

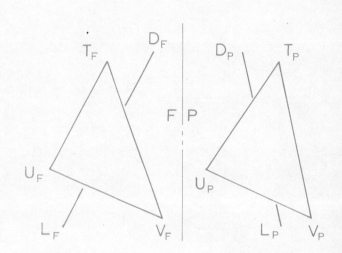

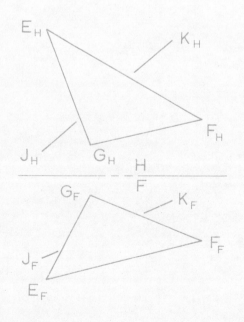

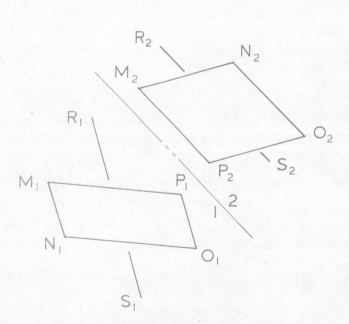

DR. NO.:
GRADE:
DR. BY:
INSTRUCTOR:
DATE:
TABLE:
COURSE:
SECTION:

18c. PROBLEM: LINE OF INTERSECTION—PLANES/EDGE VIEW METHOD

1. Solution Principle

The line of intersection between two planes is found in a projection showing one of the planes as an edge. Any adjacent view can be located by projecting piercing points where lines on the other plane pierce the edge view.

2. Use the edge view method to find the line of intersection between the two planes in each problem. Treat the planes as opaque and show proper visibility.

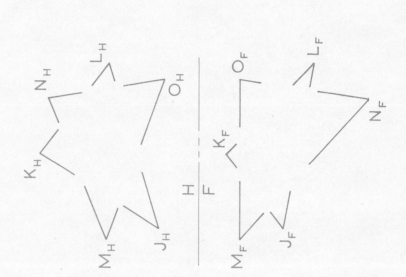

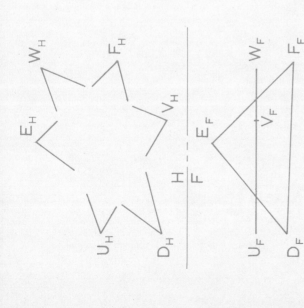

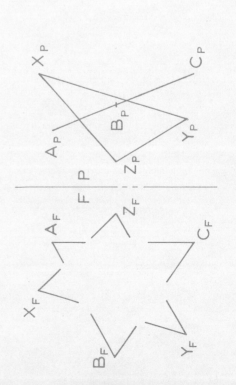

44

DR. NO.: _
GRADE:

COURSE: _
SECTION: _

DATE: _
TABLE: _

DR. BY: _
INSTRUCTOR: _

18d. PROBLEM: LINE OF INTERSECTION—PLANES/CUTTING PLANE METHOD

1. Solution Principle

An auxiliary cutting plane that intersects both given planes forms two lines of intersection that intersect at a point common to both given planes and the cutting plane.

2. Use the cutting plane method to find the line of intersection between the two planes in each problem. Treat the planes as opaque and show proper visibility.

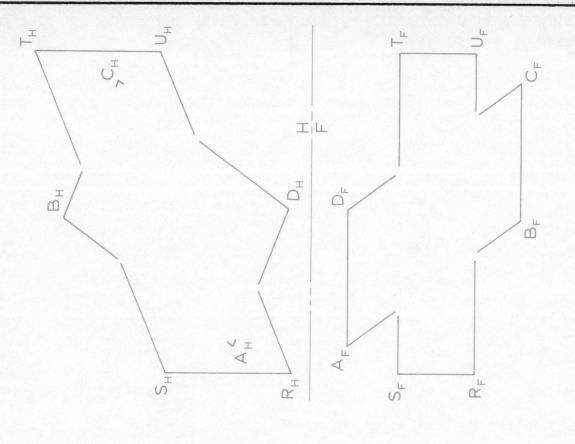

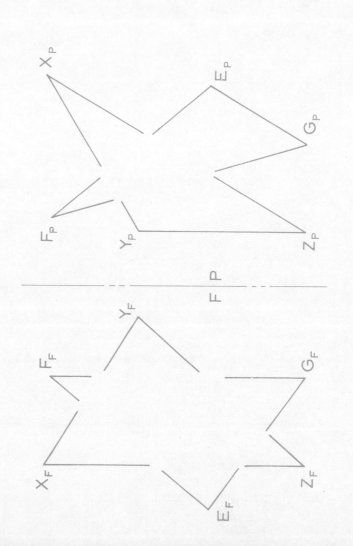

COURSE:	DATE:	DR. BY:
SECTION:	TABLE:	INSTRUCTOR:

DR. NO.

GRADE:

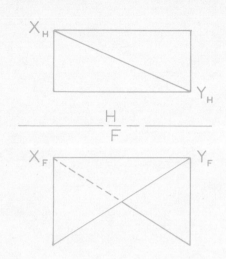

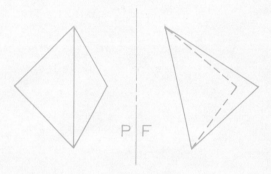

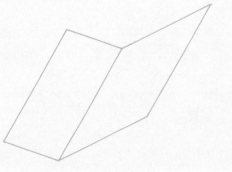

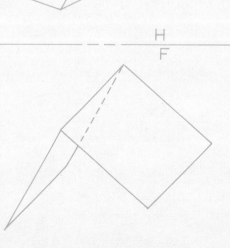

18e. PROBLEM: DIHEDRAL ANGLE

1. Solution Principle

   The dihedral angle between two planes is in true size in any view that shows the line of intersection as a point and the planes as edges.

2. Find the dihedral angle between the planes.

18f. PROBLEM: DIHEDRAL ANGLE AND LINE OF INTERSECTION

A. Find the line of intersection between the planes.

B. Find the dihedral angle between the planes.

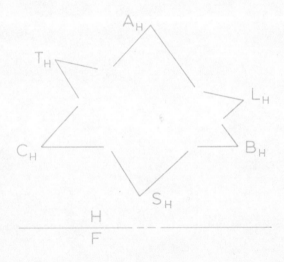

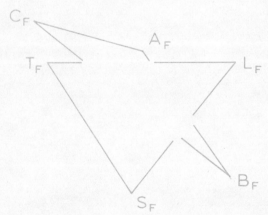

18g. PROBLEM: DIHEDRAL ANGLE APPLICATION

48

DR. NO.__
GRADE:

The tandem roller pictured here is used to compact asphalt road surfaces. Many dihedral angles are involved in the design. Projections of the bracket for the draw bar hitch are shown below. What are the dihedral angles formed by these planes?

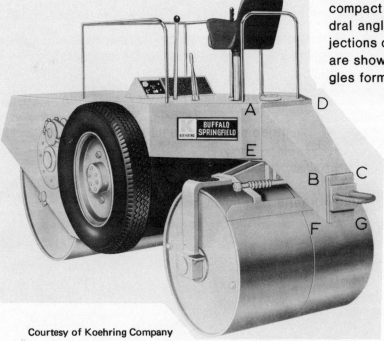

Courtesy of Koehring Company

DR. BY:
INSTRUCTOR:

DATE:
TABLE:

COURSE:
SECTION:

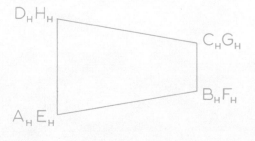

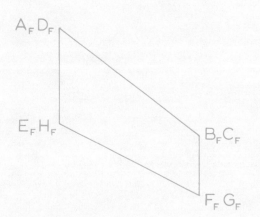

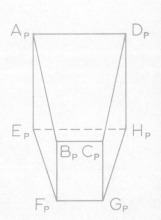

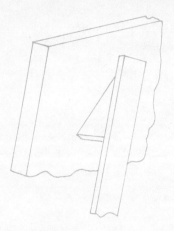

18h. PROBLEM: ANGLE BETWEEN LINE AND PLANE

1. State the solution principle for finding the true size of the angle between a line and a plane.

2. A brace located along line BR is to be welded at an angle to the plane located by points PLN. Find the true size of the angle between the line and the plane.

N_H

R_H

B_H

P_H

L_H

$\dfrac{H}{F}$

L_F

R_F

N_F

P_F

B_F

18i. PROBLEM: ANGLE BETWEEN LINE AND PLANE—DIHEDRAL ANGLE

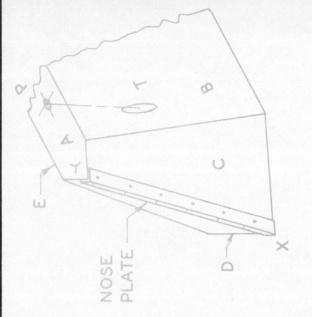

NOSE
PLATE

BRIDGE ABUTMENT

To construct the bridge abutment and fabricate the nose plate, the following information is desired:

a) A complete plan and elevation view.

b) Angle between faces A and B.

c) Angle between faces B and C.

d) Angle between faces C and D.

e) Angle between nose edge and face A.

f) Location of point T where the center line of a drain pipe through P will intersect the face of the abutment.

Note: Batter of all faces is 4:1.

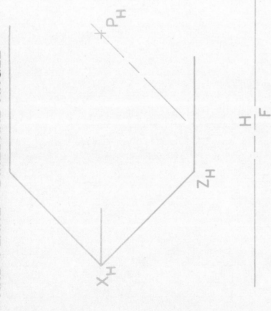

EL. 22 FT.

EL. 0 FT.

COURSE:‾	DR. BY:⹀
SECTION:‾	INSTRUCTOR:⹀
DATE:⹀	DR. No.‾
TABLE:‾	GRADE:

51

ORTHOGRAPHIC PROJECTION—SOLIDS

LEARNING GOALS

The student should:

- Increasingly develop ability to visualize shapes and forms in three dimensional space.

- Be able to produce correct orthographic views of a three dimensional object.

- Be able to provide correct visibility to lines and surfaces in orthographic views.

- Be able to select the most useful and appropriate views to describe the shape of an object.

APPLICATIONS

- Industrial designs are invariably three dimensional. Application of orthographic projection to solids is a natural extension of its application to points, lines and planes.

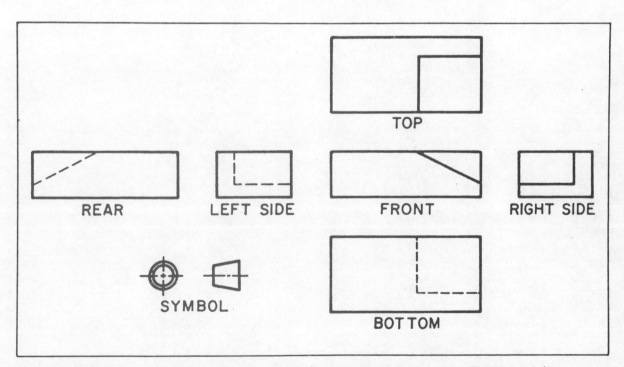

THE SIX PRINCIPAL ORTHOGRAPHIC VIEWS (Standard Arrangement—Third Angle)

19. IDENTIFICATION AND INTERPRETATION—SOLIDS

A. Identify the surfaces by placing the correct letter in each closed area in the orthographic views.

B. Complete the missing front view and identify all visible surfaces in each view.

C. Place a T or F in the circles below the right profile views to indicate correctness of each. Make a small pictorial sketch of each correctly drawn object.

D. Sketch four different but correct right profile views. Make a small pictorial sketch of each.

DR. NO._

GRADE:

COURSE:_   DATE:=   DR. BY:=

SECTION:_   TABLE:_   INSTRUCTOR:=

20. TRANSPOSING FROM PICTORIAL TO ORTHOGRAPHIC

Sketch the necessary orthographic views for complete shape description of the given objects.

A.

B.

C.

D.

E.

F.

COURSE:	DR. BY:	DR. NO.
SECTION:	INSTRUCTOR:	GRADE:
DATE:		
TABLE:		

21. COMPLETION PROBLEMS, PICTORIAL SKETCHES

Complete a third view from the given two complete views. Cross out any unnecessary views. Add a small pictorial sketch of each object.

A.

B.

C.

D.

E.

F.

DR. No.__
54
GRADE:

COURSE:__   DATE:__   DR. BY:__
SECTION:__   TABLE:__   INSTRUCTOR:__

22. APPLICATIONS—NECESSARY VIEWS

A. Prepare the necessary orthographic views of the Leader Bracket.

B. Prepare the necessary orthographic views of the Offset Mount.

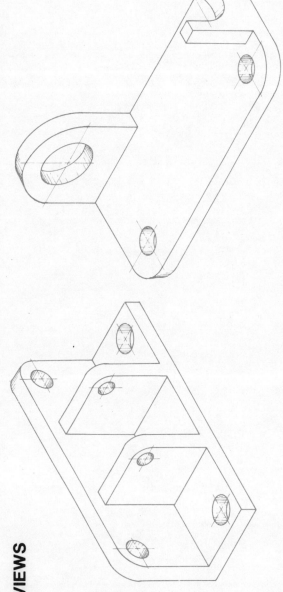

COURSE :‾‾
SECTION:‾‾

DATE: ‾‾
TABLE: ‾‾

DR. BY:‾‾
INSTRUCTOR:‾‾

DR. No.‾‾
GRADE:

23. APPLICATIONS—AUXILIARY/PARTIAL VIEWS

A. Draw the indicated auxiliary projection of the inclined surface.

B. Draw a true size view of the inclined face of the Angle Mount.

C. Draw a true size view of the right-hand end of the Slotted Angle Stop.

DR. No.
GRADE:

56

COURSE :
SECTION:

DATE :
TABLE :

DR. BY:
INSTRUCTOR:

57

SECTIONS AND CONVENTIONS

LEARNING GOALS

The student should:

- Be able to identify by name the various types of section views.

- Be able to apply the standard crosshatching symbols and cutting plane repesentation to drawings.

- Be able to apply "conventional practices" when necesssary to clarify a drawing.

- Develop ability to select the appropriate section and/or "conventional practice" to suit the conditions presented in a problem.

APPLICATIONS

- Section views and conventional practices are accepted means of clarifying orthographic views of solids. They are used to compliment the regular multi-view drawings of various industrial designs.

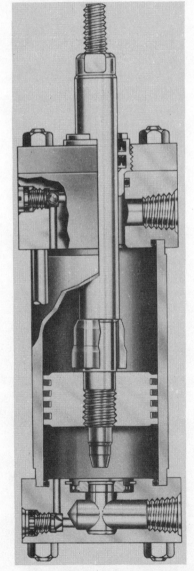

Parker Hannifin Corporation

24. VARIOUS SECTION VIEWS AND CONVENTIONS

58

A. In the space indicated complete a section view of each object as indicated by the cutting plane. Give the type of section for each.

1. Type _____

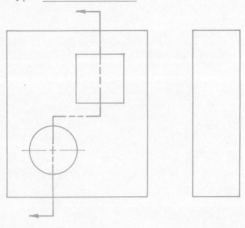

2. Type _____

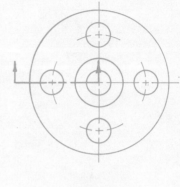

3. Type _____

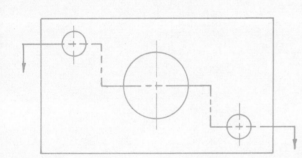

4. Type _____

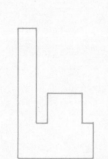

B. Project from the given horizontal and profile views to complete a view in the normal front view position. Make this front view a section view.

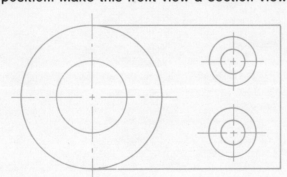

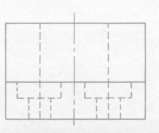

DR. NO.__
GRADE:

DR. BY:__
INSTRUCTOR:__

DATE:__
TABLE:__

COURSE:__
SECTION:__

59

25. APPLICATION PROBLEMS: SECTIONAL AND CONVENTIONAL PRACTICES

A. The view shown is a length of 1 1/2 × 1 1/2 × 3/16 (flange width × stem depth × thickness) structural tee (see structural steel shapes in text). Complete a revolved section.

B. 1. Use phantom lines to show the box lid in an alternate position.

2. Use a broken out section to clarify inner contours.

3. Use a removed section (double size) to show details of the latch.

26. APPLICATION PROBLEM: INDUSTRIAL

A. Complete the views of the Tractor Yoke, illustrating revolved and broken out sectioning.

B. Using conventional practices where applicable project a section view from the given view.

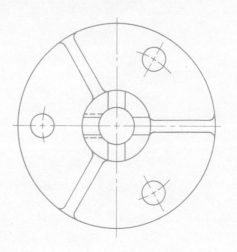

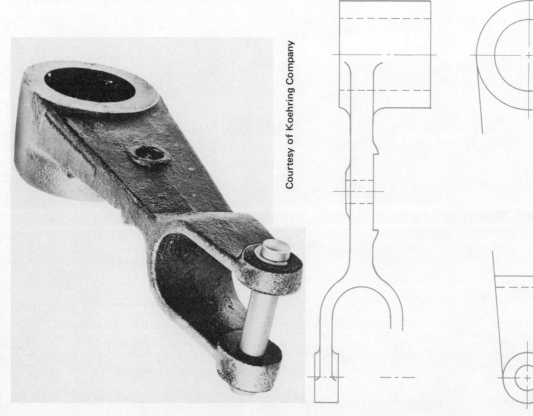

Courtesy of Koehring Company

| DR. No._ |
| GRADE: |

| DR. BY:= | DATE:= |
| INSTRUCTOR:= | TABLE:_ |

| COURSE:_ |
| SECTION:_ |

27. APPLICATION PROBLEM: ASSEMBLY SECTION

Make a two view Assembly Drawing of the Adjustable Guide. Include an offset section. Double the dimensions taken from the exploded pictorial.

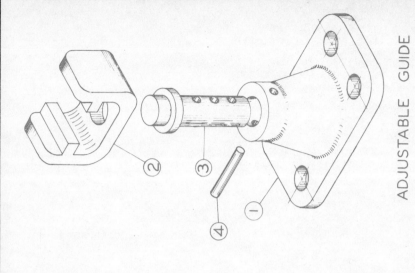

ADJUSTABLE GUIDE

COURSE:

SECTION:

DATE:

TABLE:

DR. BY:

INSTRUCTOR:

DR. No.

GRADE:

BASIC DIMENSIONING

LEARNING GOALS

The student should:

- Become familiar with the basic rules and techniques of dimensioning.

- Become familiar with the preferred dimension placement for common geometric shapes and contours.

- Develop an ability to visualize objects for the purpose of break down into basic contours for dimension selection and placement.

APPLICATIONS

- Basic dimensioning is used for size description of objects in innumerable applications.

- Manufacture and fabrication of parts involved in all areas of engineering require complete size description.

- Assembly of products and locations in many applications require dimensions.

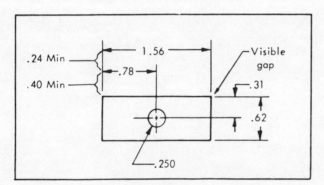

SPACING OF DIMENSIONS

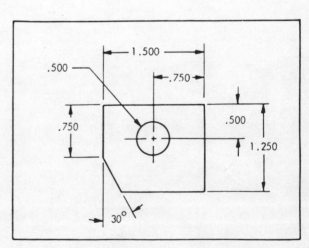

PLACEMENT OF DIMENSIONS

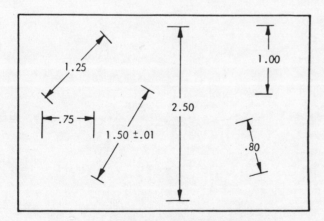

UNIDIRECTIONAL DIMENSIONS

28. APPROACH, RULES, TECHNIQUES

BASIC DIMENSIONING

A. Four basic geometric shapes that comprise the contours of many objects are shown. Dimension the size of each using the preferred positions for dimensions.

POSITIVE CYLINDER

PRISM

NEGATIVE CYLINDER

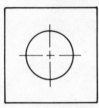 

TRUNCATED CONE

 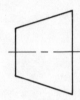

B. The following are contours that are common to many machine parts, and have standardized size dimensions. Dimension the size of each contour using standardized dimensioning.

OBJECT WITH ROUNDED ENDS

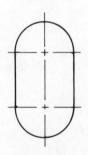 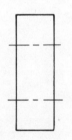

SLOTTED HOLE

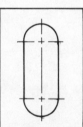 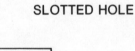 

BOLT CIRCLE

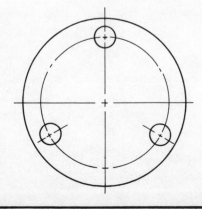 

CHAMFER

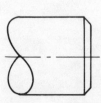 

DR. NO.:
GRADE:

DR. BY:
INSTRUCTOR:

DATE:
TABLE:

COURSE:
SECTION:

29. BASIC DIMENSIONING PROCEDURES

A. The following is a procedure for dimensioning similar to that given in the text. Follow this procedure carefully for the object shown.

1. Study the object to isolate basic geometric shapes. Identify the shapes involved.

   a. _____

   b. _____

   c. _____

   d. _____

2. Use necessary orthographic views for complete shape description. Space so there is ample room for dimensions between and around views. (This has been done for you)

3. Place size dimensions on each geometric shape identified in 1 above.

   a. Use preferred positions for size dimensions.

   b. Delay size dimensions for the contours that have preferred size dimensions in the form of a note.

4. Place location dimensions to locate contours with respect to each other.

   a. Locate circular contours by center lines.

   b. Use contour rule.

   c. Place dimensions between views or adjacent to contour (use judgement).

   d. Place shorter dimensions closer to view. Move size dimensions if necessary to achieve this.

5. Add overall dimensions.

6. Add all notes and title. Locate in convenient places.

7. Study dimensions.

   a. If any contour can be located in more than one way remove dimensions until this over-dimensioning no longer exists.

   b. Be sure that each contour (geometric shape) has a size and a location.

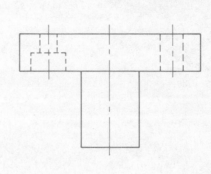

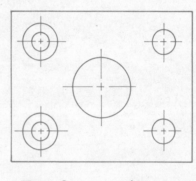

COURSE: _
SECTION: _

DATE: _
TABLE: _

DR. BY: _
INSTRUCTOR: _

DR. NO. _
GRADE:

64

30. FUNDAMENTAL APPLICATION PROBLEMS

Place extension and dimension lines in preferred positions. Add numerals by scaling the drawings, full size, in mm, and give complete information for manufacture.

A.

B.

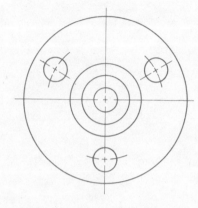

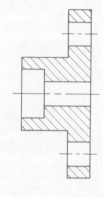

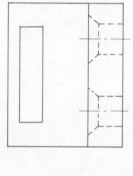

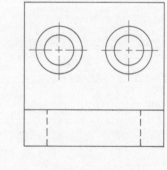

DR. No.⎯

GRADE:

DR. BY:⁼

INSTRUCTOR:⁼

DATE:⁼

TABLE:⎯

COURSE:⎯

SECTION:⎯

31. BASIC DIMENSIONING

31a. PROBLEM: INDUSTRIAL PARTS

BASIC DIMENSIONING

A. Dimension the handle and the adjustment flange each with complete information for manufacture. Scale 1:2.

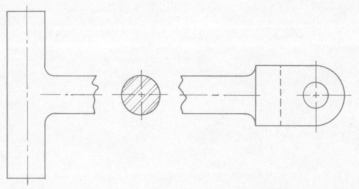

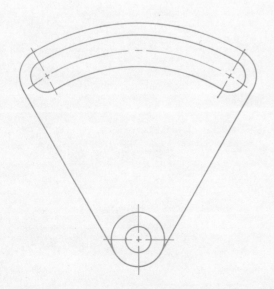

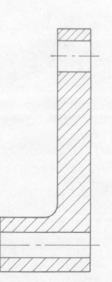

DR. BY: INSTRUCTOR:

DATE: TABLE:

COURSE: SECTION:

31b. PROBLEM: INDUSTRIAL PARTS

1. Prepare the necessary orthographic views of the Slide Stop by scaling in mm the pictorial which is shown half size.

2. Dimension the orthographic views with complete information for manufacture. The hole is 10 mm dia. with a 15 mm counterbore, 8 mm deep.

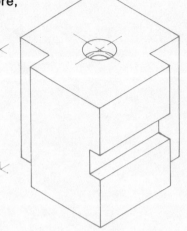

DR. NO.:
GRADE:

DR. BY:
INSTRUCTOR:

DATE:
TABLE:

COURSE:
SECTION:

PICTORIAL SYSTEMS

32. PICTORIAL DRAWING—OBLIQUE

LEARNING GOALS

The student should:

- Develop an ability to visualize objects of all shapes in three dimensions.

- Be aware of the various pictorial systems generally in use.

- Select the pictorial system that can be used to best advantage for any specific situation.

- Develop the skill necessary to produce pictorial drawings by either instrument or freehand procedures.

APPLICATIONS

- Pictorial drawings are used in many applications where quick recognition is preferred over accuracy of detail.

- Engineers utilize pictorial drawings as idea sketches, memory aids, creativity stimulation, concept development, and communication of many types of information.

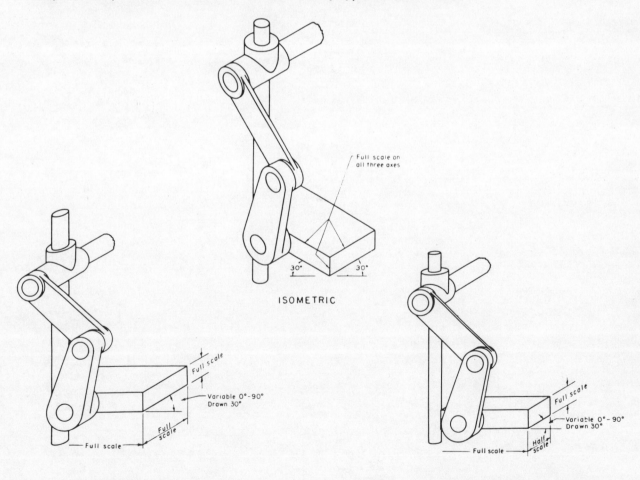

ISOMETRIC

OBLIQUE PROJECTION - CAVALIER

OBLIQUE PROJECTION - CABINET

32a. PROBLEM: OBLIQUE DRAWING

1. Make a cavalier oblique drawing, double size, of objects 1 and 2.

2. Make a cabinet oblique drawing, double size, of objects 3 and 4.

1.

2.

3.

4.

1.

2.

3.

4.

COURSE: _	DATE: =	DR. BY: =	69
SECTION: _	TABLE: _	INSTRUCTOR: =	
			DR. NO._
			GRADE:

32.b. PROBLEM: OBLIQUE DRAWING

1. Make a cavalier oblique drawing, double size, of the object shown.

2. Label to indicate the axis position used.

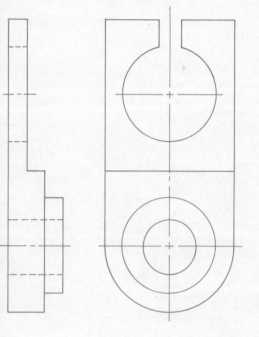

DR. No.⁻
GRADE:

DR. BY:⁼
INSTRUCTOR:⁼

DATE:⁼
TABLE:⁻

COURSE:⁻
SECTION:⁻

32.c. PROBLEM: OBLIQUE/APPLICATION

1. Make a cavalier drawing of the power brake assembly as simplified in the orthographic views.

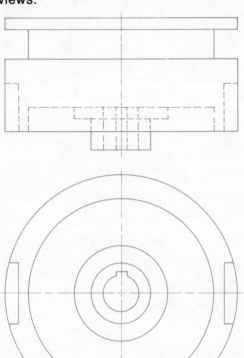

Courtesy of Simplatrol Products Corporation

DR. No.__ GRADE:

DR. BY:__ INSTRUCTOR:__

DATE:__ TABLE:__

COURSE:__ SECTION:__

33. PICTORIAL DRAWING—ISOMETRIC

33a. PROBLEM: ISOMETRIC DRAWING

1. An isometric drawing is a pictorial drawing showing the three principal axes equally spaced at 120° and with all three axis measurements at the same scale.

2. Make an isometric drawing of each of the following letters using the axes positions indicated. Use very light construction lines and do not erase.

    a. Normal axes. Full size, 30 mm depth.

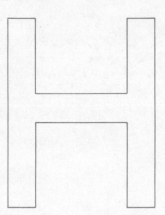

    b. Reversed axes. Full size, 30 mm depth.

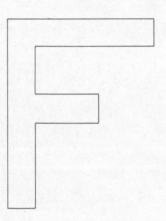

73

33b. PROBLEM: ISOMETRIC DRAWING

1. Make an isometric drawing of the object shown (Scale 3:1). Use very light construction lines and do not erase. Show 4-center construction for the ellipse.

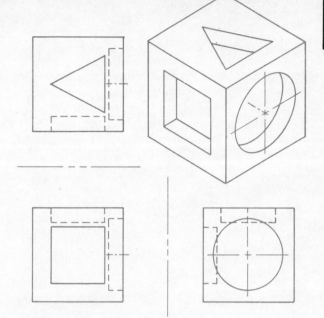

DR. NO.—
GRADE:

DR. BY:—
INSTRUCTOR:—

DATE:—
TABLE:—

COURSE:—
SECTION:—

74

33c. PROBLEM: INDUSTRIAL APPLICATION—ISOMETRIC

1. Make an isometric drawing (as assigned) of the transistor station case as simplified in the orthographic views.

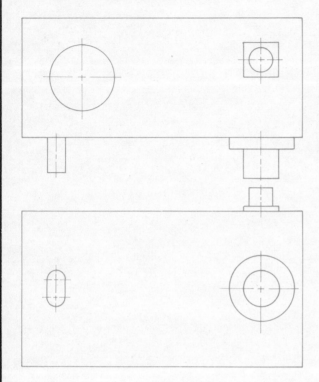

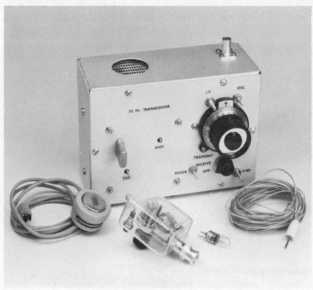

Courtesy of the American Radio Relay League, Inc.

DR. NO.

GRADE:

DR. BY:

INSTRUCTOR:

DATE:

TABLE:

COURSE:

SECTION:

75

STANDARD FASTENERS

LEARNING GOALS

A student should:

• Become familiar with the types of threaded fasteners in common use.

• Become familiar with the nomenclature associated with fasteners.

• Know the standard symbols used to represent fasteners on engineering drawings.

• Know the standard specifications for threaded fasteners.

APPLICATIONS

• Standard fasteners are used to assemble component parts together in every field of engineering.

• Although removable and permanent fasteners are used in a great variety of applications, an overwhelming majority of the applications utilize threaded fasteners.

34. THREADED FASTENERS—METRIC

A. By labeled sketches illustrate schematic and simplified thread representation, for an external thread.

B. Identify each part of the given thread specification: M 12 × 1.75C × 40

C. Sketch a flat head cap screw with the thread specification in the preceding question and indicate standard dimensions.

D. List the types of heads on American Standard and Metric Cap Screws that are meant to be:

  1. Recessed.

  2. Not recessed.

E. Parts 1 and 2 are to be fastened together by an 8 mm bolt with nut and lock washer. Assume 0.8 mm clearance where applicable.

  1. What is the diameter of the hole in part 1? _____ Part 2? _____

  2. What is the minimum standard length for the bolt? _____

  3. Write a complete specification for the bolt.

  4. Write a complete specification for the nut.

  5. Write a complete specification for the lock washer.

  6. Complete the sectioned view.

 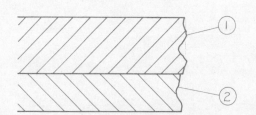

76

DR. NO.:___
GRADE:___

DR. BY:=
INSTRUCTOR:=

DATE:=
TABLE:___

COURSE:___
SECTION:___

35. THREADED FASTENERS—ENGLISH

77

DR. NO.:⎯

GRADE:

DR. BY:⎯

INSTRUCTOR:⎯

DATE:⎯

TABLE:⎯

COURSE:⎯

SECTION:⎯

A. Identify each part of the given thread specification: 3/4 — 10 UNC — 3B.

B. Write a complete specification for a 3/4 inch long, number 4, round head, machine screw having UNC threads.

C. Indicate standard dimensions of the fastener specified in the preceding question by means of a small freehand sketch.

D. Why is it impossible to recess a hex-head cap screw?

E. Parts 1 and 2 are to be fastened together by a $\frac{3}{8}$ inch recessed socket head cap screw threaded into a blind hole. Assume $\frac{1}{32}$ inch clearance where applicable.

　　1. What is the diameter of the hole in part 1? _____

　　2. What is the diameter of the tap drill used in part 2? _____

　　3. Assuming minimum thread engagement equal to the thickness of a standard hex nut, what length, using standard increments, would be required for the cap screw? _____

　　4. What is the depth of thread in part 2 assuming 4 pitches deeper than the fastener? _____

　　5. What is the tap drill depth assuming $\frac{3}{16}$ inches deeper than the thread depth? _____

　　6. Write a complete specification for the cap screw.

　　7. Complete the sectioned view.

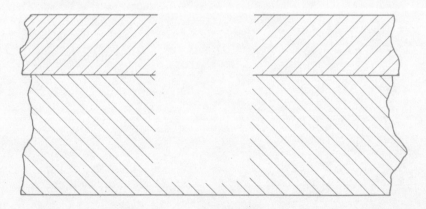

36. FASTENERS REMOVABLE AND PERMANENT

A. Keys, splines and pins are examples of _____ fasteners.

B. Riveting, soldering, brazing, gluing and welding are examples of _____ fasteners.

C. Complete the section view of a 1″ diameter shaft including a standard square key assembled into a keyseat.

D. Show a #606 Woodruff key in position to prevent motion between the shaft and pulley. Scale: 1″ = 1″

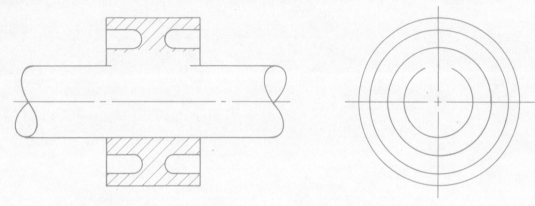

E. Apply a weld symbol indicating a 0.2″ fillet weld on both sides of the joint.

F. Complete the section view of two steel plates, fastened together by a V-groove weld. Indicate the weld by a weld symbol specifying a 0.18″ chamfer on the arrow side of the joint.

DIMENSIONING FOR PRODUCTION

LEARNING GOALS

The Student Should:

- Understand the impact of production dimensioning—the effects of imposing limits on the permissible variation in the value of a dimension.

- Be able to calculate limit dimensions and use the tables that prescribe classes of fit between mating parts.

- Learn the terminology associated with production dimensioning.

- Understand when limit dimensions are needed and when they are not necessary.

APPLICATIONS

- Limit dimensions and prescribed clearances or interferences are usually needed when mating parts occur in an assembly, such as cylinders and holes, shafts and bearings, sliding keys etc. The engineer must be familiar with all aspects of limit dimension applications in order that the product he designs will function properly.

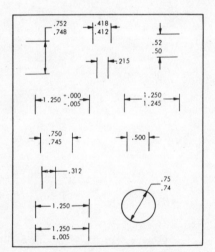

DECIMAL INCH DIMENSIONS

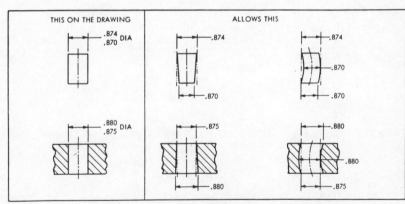

EXTREME VARIATIONS OF FORM ALLOWED BY A SIZE TOLERANCE

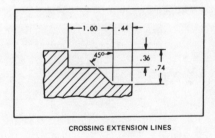

CROSSING EXTENSION LINES

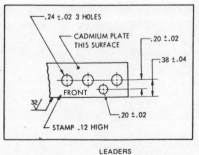

LEADERS

37. BASIC TERMS

DIMENSIONING AND PRODUCTION

Study carefully only those paragraphs in your text dealing with definitions, then close your book and check your memory and comprehension by answering the questions below. Give word answers or underline correct answers as appropriate.

1. The measurement used for general identification: _____

2. Theoretical size from which the _____ for a dimension are derived by application of allowances and tolerances: called _____ _____ .

3. The measured size of an individual part after it is fabricated: _____
   _____ .

4. Limits are the specified _____ and _____ size of a dimension.

5. That condition which leaves the most material on the part after processing: _____ .

6. *Design size* is that size to which the _____ is applied to give its _____ .
   In the *unilateral system* the design size is the (MML) (mml). In the bilateral system, the design size is the (MML) (mml) (Average material condition).

7. Tolerance is the _____ _____ _____
   in the actual size of a part or the location of some feature.

8. a. Allowance is the numerical value of the (loosest) (tightest) acceptable fit designated
      for _____ surfaces.

   b. Allowance can be *positive* (maximum) (minimum) acceptable (clearance) (interference); *negative* (maximum) (minimum) acceptable (clearance) (interference); or
      _____, (no clearance or interference).

9. A _____ system of tolerancing allows variation in two directions from a _____ size.

10. Add appropriate terms in the blanks below:

    A = Minimum shaft diameter.
    B = Maximum shaft diameter.
    C = Minimum hole diameter.
    D = Maximum hole diameter.

    D—C = _____

    D—A = _____

    C—B = _____

    B—A = _____

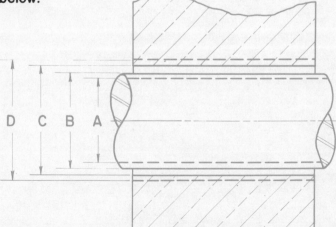

80

DR. NO.___
GRADE:

DR. BY:___
INSTRUCTOR:___

DATE:___
TABLE:___

COURSE:___
SECTION:___

38. UNILATERAL/BILATERAL SYSTEMS

COMPLETE THE FOLLOWING TABLES
English units are all in inches
Metric units are all in millimeters

A. UNILATERAL DIMENSIONING

	NOM SIZE	TOL.	ALLOW.	BASIC HOLE		BASIC SHAFT		FIT (C, T, or I)
				HOLE	SHAFT	HOLE	SHAFT	
ENGLISH	1.5	0.003	+0.004					
	2					1.997 / 1.999	2.000 / 1.998	
METRIC	20	0.006	−0.015					
	80	0.05			79.90			

B. BILATERAL DIMENSIONING

	NOM SIZE	TOL.	ALLOW.	BASIC HOLE		BASIC SHAFT		FIT (C, T, or I)
				HOLE	SHAFT	HOLE	SHAFT	
ENGLISH	1		+.001	1.000 ±.003				
						1.503 ±.003	1.500 ±.003	
METRIC	25				24.984 ±.006			
	60	0.04	−0.04					

DR. No.: ___ GRADE: ___

DR. BY: ___ INSTRUCTOR: ___

DATE: ___ TABLE: ___

COURSE: ___ SECTION: ___

39. APPLICATION OF LIMIT TABLES

Fill in the limit dimensions on the drawing and complete the table.

	A & B	C & D	E & F	G & H
NOMINAL SIZE	0.9	1.3	1.5	0.3
BASIC SIZE				
TYPE OF FIT	RC5	FN1	RC4	RC8
TOLERANCE ON HOLE				
TOLERANCE ON SHAFT				
ALLOWANCE				

COURSE: ‾ DATE: ‾ DR. BY: ‾ DR. NO. ‾

SECTION: ‾ TABLE: ‾ INSTRUCTOR: ‾ GRADE:

40. PRODUCTION DIMENSIONING: APPLICATIONS

All dimensions are in millimeters (*Show all calculations.*)

A. Complete the dimensions shown for the mating parts of a precision hinge. Where necessary to maintain an accurate fit use a +0.10 allowance and a 0.15 tolerance. Scale: 1:1

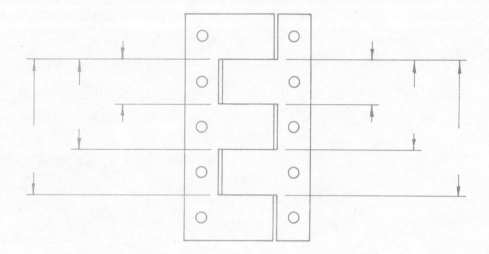

B. Complete the five dimensions indicated. The shaft diameter has a +0.066 allowance with the bracket holes and a −0.030 allowance with the pulley. Total minimum end play of the pulley is +0.180. Each tolerance is .015. Write an equation indicating the limits that determine minimum end play. Scale: 1:2

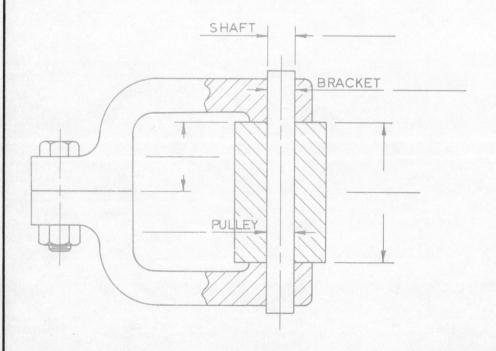

DR. NO.:

GRADE:

DR. BY:

INSTRUCTOR:

DATE:

TABLE:

COURSE:

SECTION:

41. PRODUCTION DIMENSIONING: APPLICATIONS

41a. LIMIT DIMENSIONS

A. Determine the indicated limit dimensions from the following data:

1. Shaft D has an allowance of +0.040 with bushing hole F and −0.035 with pulley hole E. Tolerance is 0.015. Nominal diameter is 25 mm. Use basic shaft system.

2. Bushing H has an allowance of −0.035 with bracket hole G. Tolerance is 0.015. Nominal diameter is 35 mm. Use basic shaft system.

3. Pulley length C is to have a minimum end play of +0.16 with the interior faces of the brackets. Nominal A is 180 mm. Nominal B is 8 mm. Tolerance on all dimensions is 0.05. Use basic hole system.

Show computations in an orderly manner. All dimensions are in millimeters.

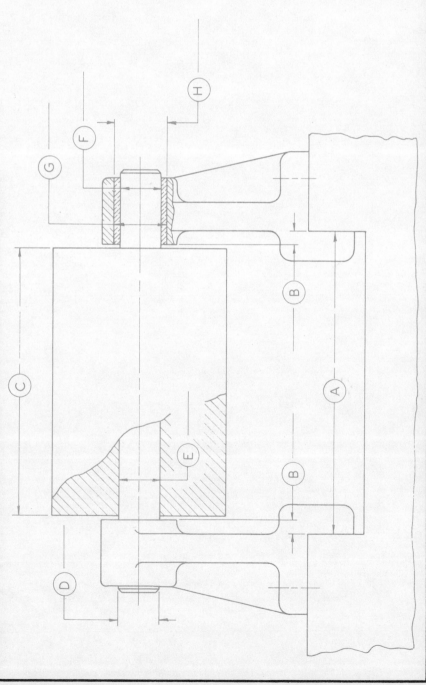

COURSE:
SECTION:

DATE:
TABLE:

DR. BY:
INSTRUCTOR:

41b. PRODUCTION DIMENSIONING: APPLICATIONS

The section drawn below shows the mounting of an Idler Shaft in a Sprocket Assembly. Detail drawings of the parts involved are drawn and partially dimensioned on page 86. Specifications concerning the fits between mating parts are in the table. Complete the dimensioning according to these specifications.

All work is to be done on page 86. Calculations should be done on the back of the sheet.

SPEC	PART	NOM. SIZE	TOL.	NOTES
A	5,7	1.2		FN2
B	3,5	.9		RC4
C	6,7	.5		LC4
D	3	4.6	.002	Min end play of
D	5	.2	.002	shaft +0.005*
D	6	5	.002	5.000 Max
E	1,3	1.5		LC2
F	2,3	1.5		LC2
G	3,4	1.5		LC2
H	1	1	.002	Total
H	2	1.1	.002	Allow
H	4	2.5	.002	+0.002**
J	1,3			No. 406
J	2,3			Woodruff Key
K	3			.1 X .1 Chamfer
L	6			.05 X .05 Chamfer
M	6			1/2-13 UNC

*Write an equation indicating the limits involved at minimum end play of the shaft and show calculations for the dimensions on part 3.

**The sum of the lengths of parts 1, 2 and 4 must be less than dimension D on part 3. Write an equation indicating the limits involved at the closest condition and calculate the length of the spacer.

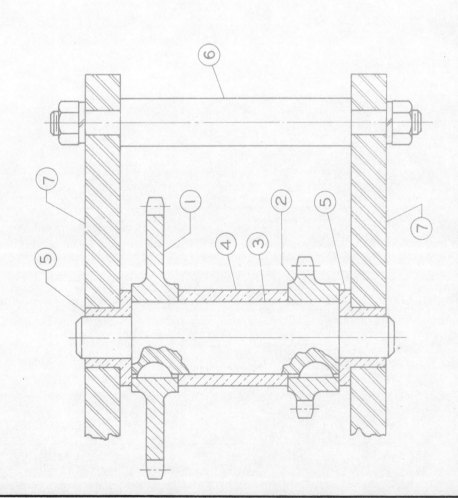

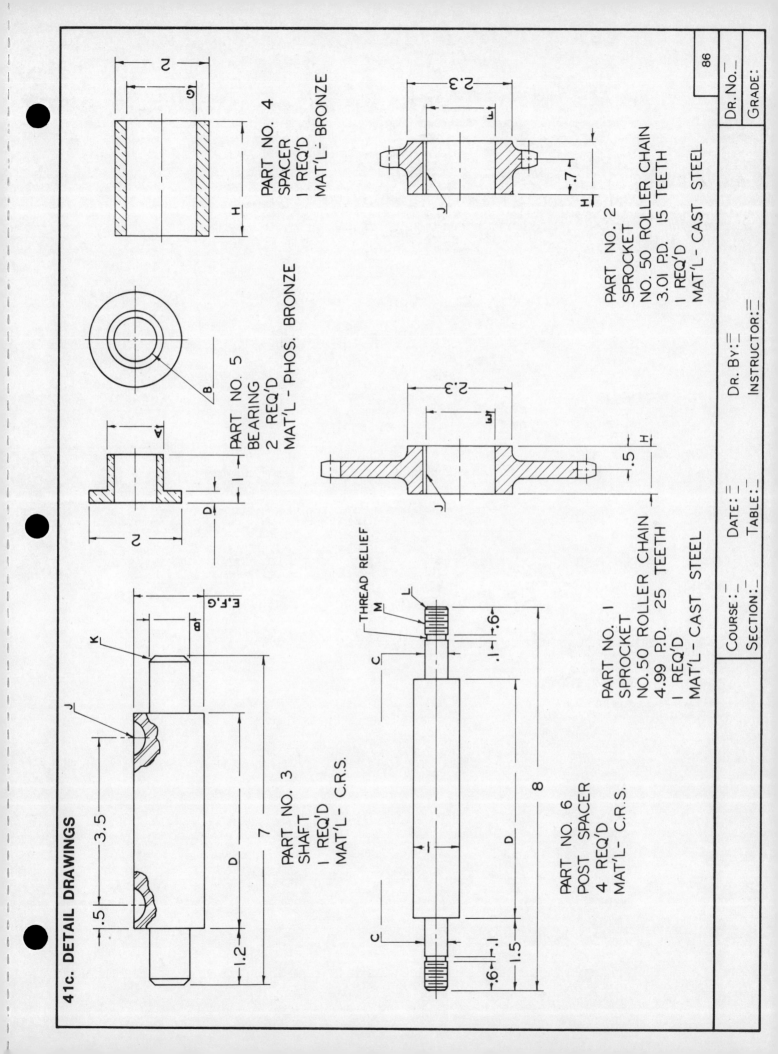

41c. DETAIL DRAWINGS

PART NO. 4
SPACER
1 REQ'D
MAT'L - BRONZE

PART NO. 5
BEARING
2 REQ'D
MAT'L - PHOS. BRONZE

PART NO. 3
SHAFT
1 REQ'D
MAT'L - C.R.S.

PART NO. 2
SPROCKET
NO. 50 ROLLER CHAIN
3.01 P.D.   15 TEETH
1 REQ'D
MAT'L - CAST STEEL

PART NO. 1
SPROCKET
NO. 50 ROLLER CHAIN
4.99 P.D.   25 TEETH
1 REQ'D
MAT'L - CAST STEEL

THREAD RELIEF

PART NO. 6
POST SPACER
4 REQ'D
MAT'L - C.R.S.

49

DR. NO.:____

GRADE:

DR. BY:____

INSTRUCTOR:____

COURSE:____   DATE:____

SECTION:____   TABLE:____

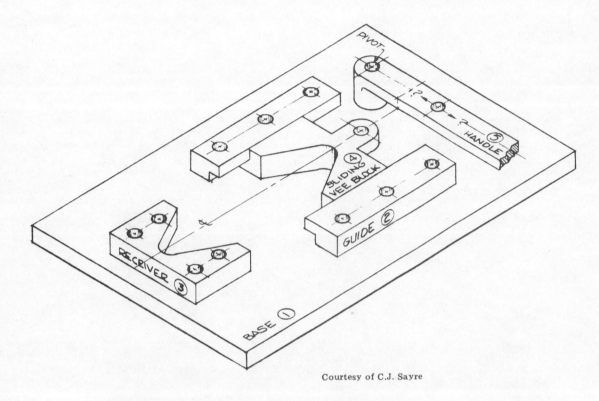

Courtesy of C.J. Sayre

A clamping device is shown in pictorial to a scale of 1″ = 4″. There are two options for the student to choose from in this project. Both involve making two complete and well-executed drawings. The pictorial should be scaled to the nearest .1″. Nominal (non-critical) dimensions shall be specified to two decimal places. Dimensions affecting fit shall be specified to four decimal places, as given in ANSI B4.1 tables.

Option I: By carefully scaling dimensions from the illustration, make good complete drawing of part ② and ④ . The sliding block must have a class RC3 fit in its guides in order for it to slide smoothly. For purposes of dimensioning it will be assumed that the guides, part ② , are fixed to the baseplate by means of pins (Dia. = .5000 ± .0002) which have an FN 2 fit with both the base plate ① and parts ② . It is also assumed that the base plate holes are precisely 6.0000 center to center distance.

Option II: Design an operating link to connect the handle ⑤ to the sliding block ④ . It is necessary to assure that the slider shall move forward exactly 3 inches as the handle rotates 30° clockwise from its illustrated position. The end holes of the link shall have an RC 4 fit on standard pins measuring .5000 ± .0002 diameter. Both the layout and link detail drawings are required. Lengths of both the link member and the handle radius shall be identified clearly and dimensioned properly on the layout drawing.

DR. BY: __

INSTRUCTOR: __

DATE: __

TABLE: __

COURSE: __

SECTION: __

43. APPLICATION: PROJECT—ELECTRICAL

DR. NO.:
GRADE:
DR. BY:
INSTRUCTOR:
DATE:
TABLE:
COURSE:
SECTION:

88

An experimental design for a new 15″ Woofer hi-fi speaker is proposed with the following specifications.

Woofer cone material: 0.125 spun fiber glass
        15″ outside diameter
        4″ cone depth

Solenoid: Material is Fiberglass with length and inside diameter shown below. Wall thickness $= \dfrac{0.050}{0.048}$

     400 turns of #20 magnet Formar covered wire are to be wound in four (4) layers as shown below. Wire dia. = 0.032″ (max.)

Magnet: Material is Alnico V, overall nominal dimensions are 2.5D × 6.0

With these basic specifications given determine and draw as assigned:

a. Detail Solenoid computing "W" and "D".

b. Detail Magnet such that solenoid will fit into cylinder with .010 clearance to all surfaces.

c. Assembly drawing of cone, solenoid and magnet to allow 1″ axial motion of cone/solenoid in magnet.

d. Design frame to hold parts.

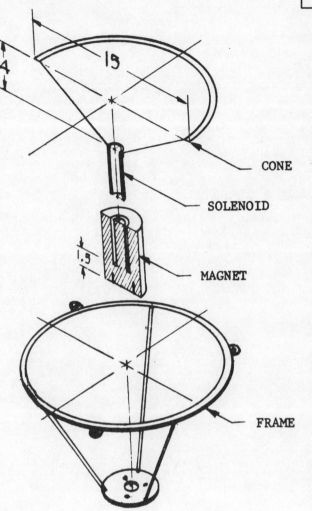

CONE

SOLENOID

MAGNET

FRAME

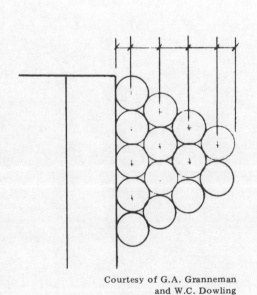

Courtesy of G.A. Granneman
and W.C. Dowling

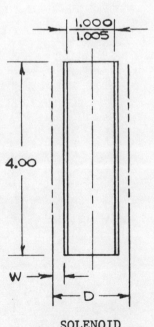

SOLENOID
CROSS SECTION
HALF SIZE

44. APPLICATION: PROJECT—PROCESSES

89

One method for disposing of ashes from a power plant involves mixing the ashes with water to form a slurry. The slurry is then transported by pipeline to a shallow man-made pond, where the water is evaporated.

The power plant shown in the sketch below must install this type of system. The pipeline is 8″ in diameter, is covered by 2″ of insulation, and is 10′ above ground. The plant has in storage 500 pieces of lumber which are 4″ × 4″ × 16′, which it recovered from a salvage operation last year. Utilizing this lumber, design a brace for the pipeline, which will be placed along the pipeline at 10′ c-c.

In addition to the detail drawing of the brace design, make a layout drawing of the sketch below, using a suitable scale, and a bill of materials for the project. Include an estimate of cost using information provided by your instructor.

COST ESTIMATES

Pipe	$12.00/ln. ft.
Insulation	$ 0.15/sq. ft.
Bracing	no cost
Fasteners	$ 0.50 each

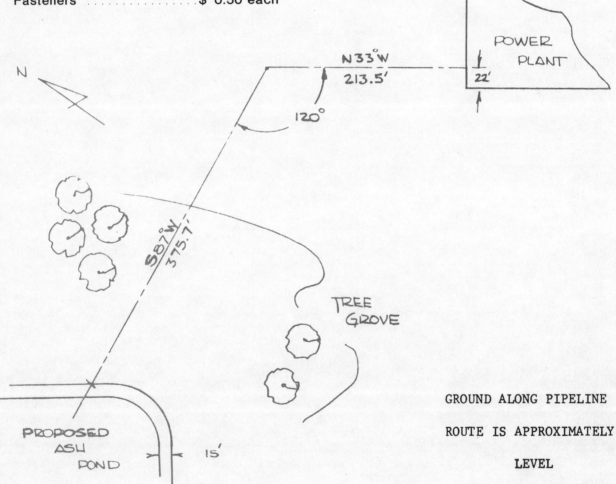

GROUND ALONG PIPELINE

ROUTE IS APPROXIMATELY

LEVEL

Courtesy of B.L. Butterfield

DR. No.:⹀ GRADE:

DR. BY:⹀ INSTRUCTOR:⹀

DATE:⹀ TABLE:⹀

COURSE:⹀ SECTION:⹀